Many Degrees of Freedom in Field Theory

NATO ADVANCED STUDY INSTITUTES SERIES

A series of edited volumes comprising multifaceted studies of contemporary scientific issues by some of the best scientific minds in the world, assembled in cooperation with NATO Scientific Affairs Division.

Series B: Physics

RECENT VOLUMES IN THIS SERIES

Volume 23 – Photon Correlation Spectroscopy and Velocimetry
edited by E. R. Pike and H. Z. Cummins

Volume 24 – Electrons in Finite and Infinite Structures
edited by P. Phariseau and L. Scheire

Volume 25 – Chemistry and Physics of One-Dimensional Metals
edited by Heimo J. Keller

Volume 26 – New Developments in Quantum Field Theory and Statistical Mechanics–Cargèse 1976
edited by Maurice Lévy and Pronob Mitter

Volume 27 – Topics in Theoretical and Experimental Gravitation Physics
edited by V. De Sabbata and J. Weber

Volume 28 – Material Characterization Using Ion Beams
edited by J. P. Thomas and A. Cachard

Volume 29 – Electron–Phonon Interactions and Phase Transitions
edited by Tormod Riste

Volume 30 – Many Degrees of Freedom in Field Theory
edited by L. Streit

Volume 31 – Many Degrees of Freedom in Particle Theory
edited by H. Satz

Volume 32 – Electron and Ion Spectroscopy of Solids
edited by L. Fiermans, J. Vennik, and W. Dekeyser

Volume 33 – Microscopic Structure and Dynamics of Liquids
edited by J. Dupuy and A. J. Dianoux

The series is published by an international board of publishers in conjunction with NATO Scientific Affairs Division

A	Life Sciences	Plenum Publishing Corporation
B	Physics	New York and London
C	Mathematical and Physical Sciences	D. Reidel Publishing Company Dordrecht and Boston
D	Behavioral and Social Sciences	Sijthoff International Publishing Company Leiden
E	Applied Sciences	Noordhoff International Publishing Leiden

Many Degrees of Freedom in Field Theory

Edited by

L. Streit
University of Bielefeld, Federal Republic of Germany

PLENUM PRESS • NEW YORK AND LONDON
Published in cooperation with NATO Scientific Affairs Division

Library of Congress Cataloging in Publication Data

International Summer Institute on Theoretical Physics, 8th, University of Bielefeld, 1976.
Many degrees of freedom in field theory.

(NATO advanced study institutes series: Series B, Physics; v. 30)
"Published in cooperation with NATO Scientific Affairs Division."
Includes index.
1. Degree of freedom–Congresses. 2. Quantum field theory–Congresses. I. Streit, Ludwig, 1938- II. Title. III. Series.
QC174.52.D43I57 1976 530.1'43 77-29217

Proceedings of the 1976 International Summer Institute of Theoretical Physics held at the University of Bielefeld, Federal Republic of Germany, August 23–September 4, 1976, published in two volumes, of which this is the first

MyCopy version of the original edition 1978
A Division of Plenum Publishing Corporation
227 West 17th Street, New York, N.Y. 10011

DOI 10.1007/978-1-4615-8924-2
www.springer.com/mycopy

Preface

Volumes 30 and 31 of this series, dealing with "Many Degrees of Freedom," contain the proceedings of the 1976 International Summer Institute of Theoretical Physics, held at the University of Bielefeld from August 23 to September 4, 1976. This institute was the eighth in a series of summer schools devoted to particle physics and organized by universities and research institutes in the Federal Republic of Germany.

Many degrees of freedom and collective phenomena play a critical role in the description and understanding of elementary particles. The lectures in this volume were intended to display how these structures occur in various recent developments of mathematical physics. Lectures ranged from classical nonlinear field theory over classical soliton models, constructive quantum field theory with soliton solutions and gauge models to the recent unified description of renormalization group techniques in probabilistic language and to quantum statistical dynamics in terms of derivations.

The Institute took place at the Center for Interdisciplinary Research of the University of Bielefeld. On behalf of all participants, it is a pleasure to thank the officials and the administration of the Center for their cooperation and help before and during the Institute. Special thanks go to V.C. Fulland, M. Kämper, and A. Kottenkamp for their rapid and competent preparation of the manuscripts.

The Institute was sponsored by the NATO Advanced Study Institute Programme and supported by the Bundesminister für Wissenschaft und Forschung of the Land Nordrhein-Westfalen. Last, but certainly not least, the valuable help of I. Andric, V. Enss, F. Jegerlehner, B. Petersson and P. Stichel in organizing the Institute is gratefully acknowledged.

March, 1977

L. Streit

Contents

AN INTRODUCTION TO SOME TOPICS IN CONSTRUCTIVE QUANTUM FIELD THEORY

Jürg Fröhlich *

Department of Mathematics

Princeton University, Princeton, N.J. 08540,USA

SUMMARY

We give an elementary introduction to the spirit and some of the simpler techniques of constructive quantum field theory. In the first part of these notes we briefly review the general framework of relativistic quantum field theory, its Euclidean description and Euclidean field theory. In the second part we combine Euclidean with Lagrangean field theory and, as a result, formulate a concrete program for the construction of relativistic quantum field models. In the third part we exemplify this program by considering in some detail the famous $\lambda\phi^4$-theory in two space-time dimensions and commenting on the construction of $\lambda\phi^4$ in three dimensions. Finally, in part four, we discribe some recent results concerning the degeneracy of the physical vacuum (phase transitions), the spontaneous breaking of internal symmetries and the critical point in $\lambda\phi^4$ in two and three dimensions.

* Supported by ZiF, University of Bielefeld, 48 Bielefeld Germany, and in part by the U.S. National Science Foundation under grant MPS 75 - 11864; A.P. Sloan Foundation Fellow

INTRODUCTION

What is Constructive Quantum Field Theory (c.q.f.t.)?

1. Orientation and definition: c.q.f.t. is the most recent branch of a long standing attempt towards achieving a mathematically rigorous understanding of relativistic quantum field theory. c.q.f.t. is concerned with the construction of explicit models of non-trivial relativistic quantum fields. It appears to be the natural continuation of axiomatic field theory.

We recall that relativistic quantum field theory (r.q.f.t.) is the attempt to combine quantum mechanics with the special theory of relativity in the form of a field theory in such a way that a realistic theory of elementary particles results. It follows from this definition that r.q.f.t. "marries" the two fundamental constants $\hbar$ and c. It is well known that this marriage has been plagued with many difficulties over a period of almost fifty years; its many successes have kept it alive.

In the course of the history of r.q.f.t. it was found useful (just because of all those difficulties) to formulate general postulates - axioms - every r.q.f.t. that deserves this name ought to satisfy and then to draw mathematically rigorous conclusions from these postulates. This enterprise has been called axiomatic field theory, [SW,J]. It has supplied a deep insight into the structure of r.q.f.t.. In spite of its many successes (PCT theorem, connection between spin and statistics, Haag-Ruelle and LSZ scattering theory, dispersion relations, etc.) axiomatic field theory has not provided a proof that the general postulates (various sets of axioms) are compatible with the obviously important requirement that the scattering matrix be different from the identity.

c.q.f.t. has grown out of this (somewhat unsatisfactory) situation and proposes to establish this compatibility by constructing model theories with a non-trivial scattering matrix fulfilling all the postulates.

So far the program of c.q.f.t. has been carried out successfully in space-times of dimension two and three; (the so called "super-renormalizable" models are under control). In four dimensions new insight into the basic difficulties of r.q.f.t. has been gained, but the constructive program is still incomplete. (see e.g. [Bi]). This is no reason to discredit c.q.f.t.. Even if it were not able to supply non-trivial models in four space-time dimensions it would nevertheless be a successful branch of mathematical physics: c.q.f.t. has lead to a new understanding of the basic problems of r.q.f.t., has brought new concepts, has supplied models that have been analyzed in great detail, has motivated some research in pure mathematics and has shown that a final success of r.q.f.t. must necessarily be the consequence of very hard and technical analysis.

2. Some comments on the history of c.q.f.t.: There are many important, less and more recent accounts of the history of this subject to which we should like to refer the reader [GJ1,2,GJS1,He,S1,W1,O].

At this point we only want to mention the names of those who have made the most substantial contributions to c.q.f.t..

At an early stage of the development of c.q.f.t. (1964-70) most of the important results were proven by J. Glimm, A. Jaffe and E. Nelson; see e.g. [GJ1,2] and refs. given there, [N1]. During this period they applied the Hamiltonian formalism; (construction of models by means of constructing Hamiltonians, quantum fields and physical vacua without cutoffs; for this see [GJ1,2,4,5, He]). For references to Segal's contributions see e.g. [S1]. Their work was paralleled by the work of K. Symanzik who applied the Euclidean formalism, (1964-69). See [Sy 1,2,3]. He proposed to construct Euclidean (imaginary time) Green's functions (by means of path space techniques and functional integrals) from which a relativistic quantum field theory could be reconstructed by analytic continuation. (The drawback of his methods was that they were more heuristic and that he did not show under what general conditions Euclidean Green's functions determined a r.q.f.t.. His methods were however of great conceptual importance). Path space- and "imaginary time" techniques were also applied in work of Glimm and Jaffe.

A very important synthesis of the Hamiltonian and Euclidean formalism was revealed and applied to c.q.f.t. by E. Nelson [N2,3], in 1971. A short time later striking applications of this synthesis were presented by F. Guerra [Gu] and by J. Glimm and A. Jaffe [GJ3].

These contributions lead to a deep understanding of the Euclidean description of r.q.f.t. (see [OS]) and its usefulness for the purposes of c.q.f.t.; (see [E,S1]).

Since 1972 all substantial results in c.q.f.t. dealing with relativistic quantum field models have been formulated and proven within the Euclidean framework. Very important results have been contributed by Glimm and Jaffe [GJ3] (see also [MS]), Glimm, Jaffe and Spencer [GJS 2,3,4] and refs. given there, Nelson [N4], Guerra, Rosen and Simon [GRS 1,2,3] (see also [S2,Nw, GRS4]), Osterwalder and Schrader [OS] (see also [Gl, F1]) and many others, (e.g. [S≠I], [$P(\phi)_2$], [Y_2], [ϕ^4_3], [FSS], etc.).

New concepts and foundational improvements have been developped hand in hand with new mathematical techniques proofs of new types of estimates, the analysis of new models throughout the history of c.q.f.t.. This interplay of ideas of different nature and from different fields in mathematics and physics and the various proposed ways of approaching c.q.f.t. have made it a fascinating subject.

3. A warning: These written up lecture notes contain nothing new. At best they may open a compact synthetic view at the subject and they may set a few novel accents. Foundational topics are probably overemphasized , technical aspects underemphasized . These notes will certainly bore the experts. I hope they are not completely useless.

I. RELATIVISTIC QUANTUM FIELD THEORY; WIGHTMAN DISTRIBUTIONS, EUCLIDEAN GREEN'S FUNCTIONS AND EUCLIDEAN FIELD THEORY, OPERATOR ALGEBRAS

I.1 Hilbert Space Formulation of r.q.f.t., [SW,J]:

Here we briefly review the Wightman axioms for a neutral, scalar quantum field ϕ in d space-time dimensions. As we mentioned already, r.q.f.t. intends to combine quantum mechanics with special relativity in the form of a field theory. It inherits

- from quantum mechanics

(Wo) The states of a physical system are the unit rays of a separable Hilbert space $\mathcal{H}$, (more generally of a family of such Hilbert spaces: the super-selection sectors).

- from field theory

(Wi) To each test function f in the Schwartz space $\mathcal{S}(\mathbb{R}^d)$ there is associated an unbounded operator $\phi(f)$ on $\mathcal{H}$ with $[\phi(f)]^* \supseteq \phi(\bar{f})$; $\phi(f)$ is defined on and leaves invariant some dense domain D (independent of f).

- from special relativity

(Wii) Covariance: There is a continuous, unitary representation $U:(a,\Lambda)\in\mathcal{P}\rightarrow U(a,\Lambda)$ of the Poincaré group $\mathcal{P}$ on $\mathcal{H}$ with the property that $U(a,\Lambda)\phi(f)U(a,\Lambda)^{*}=\phi(f_{(a,\Lambda)})$,[1) (and $U(a,\Lambda)D\subseteq D$).

- from the "stability" of relativistic systems at zero density and temperature

(Wiii) <u>Spectrum condition and vacuum</u>: The spectrum of (H,P) (the infinitesimal generators of the translations $U(a,\mathbb{1})$) is contained in the forward light cone $\bar{V}_{+}$, and $(0,0)$ is an eigenvalue of (H,P).

- from special relativity, causality and quantum mechanics (Wiv) <u>Locality</u>:

$$\phi(f)\phi(g)-\phi(g)\phi(f) = 0$$

if the supports of f and g are space-like separated, (as an operator equation on D).

- from "everything can be built up out of functions of ϕ" (Wv) <u>"Completeness"</u>: Let $\mathcal{P}(\phi)$ be the *algebra of polynomials in $\{\mathbb{1}\,\phi(f):f\in\mathcal{S}(\mathbb{R}^{d})\}$. Then there is a vector Ω in the eigenspace of (H,P) corresponding to the eigenvalue $(0,0)$ such that $\Omega\in D$, and $\{\mathcal{P}(\phi)\Omega\}$ is dense in $\mathcal{H}$.

(Wo)-(Wv) are called <u>(Gårding-) Wightman axioms</u>. For a more precise formulation and discussion see [SW,J,S1].

I.2 Wightman Distributions, [SW,J]:

It has been shown by Wightman (Wightman's reconstruction theorem) that a r.q.f.t. satisfying (Wo)-(Wv) is completely determined by the vacuum expectation values (v.e.v.'s) of all products of fields:

$$W_n(x_1,\ldots,x_n) \equiv \langle\Omega,\phi(x_1)\ldots\phi(x_n)\Omega\rangle,\ n=0,1,2,\ldots$$

(to be interpreted in the sense of distributions). Let $\underline{\mathcal{S}}$ be the family of sequences

$$\underline{f} \equiv \{f_n\}_{n=0}^{\infty} \text{ with } f_n\in\mathcal{S}(\mathbb{R}^{dn}),$$

$$f_n = 0, \text{ for all } n>n_0(\underline{f}).$$

We define

$$\underline{f}^{*} = \{f_n^{*}\}_{n=0}^{\infty},\ f_n^{*}(x_1,\ldots,x_n) = \overline{f_n(x_n,\ldots,x_1)},$$

$$\underline{f}\otimes\underline{g} = \{\sum_{k=0}^{n} f_{n-k}\otimes g_k\}_{n=0}^{\infty}$$

---

1) $f_{(a,\Lambda)}(x) = f(\Lambda^{-1}(x-a))$.

The definition and these operations make $\mathscr{S}$ a topological *algebra, the so called Borchers algebra; see e.g. [J]. Wightman has shown [2] equivalence of (Wo)-(Wv) and the following (WO)-(W4):

(WO) $W_o = 1$ ($\Longleftrightarrow$ $\langle\Omega,\Omega\rangle=1$);

temperedness: For all $n\geq 1$, W_n is a tempered distribution (i.e. $\underline{W} = \{W_n\}_{n=o}^{\infty}$ is a continuous, linear functional on $\mathscr{S}$: $\underline{W}(\underline{f}) = \sum_{n=o}^{\infty} W_n(f_n)$).

(W1) Positivity: $W(\underline{f}^*\otimes\underline{f})>0$
($\Longleftrightarrow$ scalar product of $\mathcal{H}$ is positive definite).

(W2) Invariance (covariance):
$\underline{W}(\underline{f}) = \underline{W}(\underline{f}_{(a,\Lambda)})$, for all (a,Λ) in $\mathcal{P}$.
($\Longleftrightarrow$ covariance of ϕ + invariance of Ω).

(W3) Spectrum:
A condition on supp $\tilde{W}_n$ expressing spec$(H,P)\subseteq\bar{V}_+$;
(here $f \to \tilde{f}$ denotes Fourier transformation).

(W4) Locality:
Expresses (Wiv), i.e., for $(x_i-x_{i+1})^2<0$,
$W_n(\ldots x_i,x_{i+1}\ldots) = W_n(\ldots x_{i+1},x_i\ldots)$, for all n.

An example: the free field
The free, neutral, scalar field has the following Wightman distributions:

$$W_{2n+1} = 0,\ n=0,1,2,\ldots\ .$$

$$W_2(x_1,x_2) = (2\pi)^{-d+1}\int dp\ \theta(p^o)\,\delta(p^2-m^2)\,e^{ip(x_2-x_1)}$$

$$= (2\pi)^{-d+1}\int \frac{d\vec{p}}{2\omega(\vec{p})}\ e^{i(\omega(\vec{p})(x_2^o-x_1^o)-\vec{p}(\vec{x}_2-\vec{x}_1))} \tag{I.1}$$

with $\omega(\vec{p})=\sqrt{\vec{p}^2+m^2}$ and $m>0$. Finally

$$W_{2n}(x_1,\ldots,x_{2n}) = \sum_{\text{pairings}} W_2(x_{i_1},x_{j_1})\ldots W_2(x_{i_n},x_{j_n}), \tag{I.2}$$

2) Wightman's reconstruction theorem.

with $i_1<\ldots<i_n$, $i_1<j_1,\ldots,i_n<j_n$, and $(i_1,j_1,\ldots,i_n,j_n)$ a permutation of $(1,\ldots,2n)$.

Exercise 1: Show that the W_n's defined in (I.1)-(I.2) satisfy (WO)-(W4). Needless to say that the scattering matrix S of the free field is of course equal to $\mathbb{1}$.

I.3 Are (WO)-(W4) Compatible with $S\neq\mathbb{1}$?

One of the major goals of c.q.f.t. is to prove that (WO)-(W4) are compatible with $S\neq\mathbb{1}$. In c.q.f.t. one proposes to give such a proof by constructing models with $S\neq\mathbb{1}$ fulfilling (WO)-(W4). (Of course, once such models are constructed, c.q.f.t. intends to analyze their relevance to physics).

So far all the models that have been studied in c.q.f.t. are Lagrangean r.q.f.t.'s.

Most of the rigorous work in the (Hamiltonian) period 1964-1970 was conceptually inspired by the Hilbert space formulation (Wo)-(Wv) of the Wightman axioms and the algebraic framework of Haag and Kastler [HK]; see [GJ4,5]. It lead to the experience that it is very difficult to directly construct the Wightman distributions $\underline{W}$, (e.g. by means of a direct analysis of $\mathcal{H}$,(H,P),ϕ,Ω). Since 1971 it has been realized [GJS2,N4,GRS3] that it is much easier to first construct the Euclidean Green's functions (EGF's; = W_n's restricted to imaginary times) of an r.q.f.t. Formally EGF's are defined as follows:

$$"W_n(x_1,\ldots,x_n) = \langle\Omega, \prod_{j=1}^{n} e^{it_jH}\,\phi(0,\vec{x}_j)e^{-it_jH}\Omega\rangle". \tag{I.3}$$

Assume now that $t_1<t_2<\ldots<t_n$. Axiom (W3) guarantees that $H\geq 0$. Therefore t_j may be replaced in (I.3) by it_j, yielding

$$"W_n(it_1,\vec{x}_1,\ldots,it_n,\vec{x}_n)$$

$$= \langle\Omega,\phi(0,\vec{x}_1)e^{-(t_2-t_1)H}\ldots e^{-(t_n-t_{n-1})H}\phi(0,\vec{x}_n)\Omega\rangle$$

$$\equiv S_n(x_1,\ldots,x_n)" \tag{I.4}$$

The S_n's are called Euclidean Green's or Schwinger

functions.

That it may in general be easier to construct and analyze EGF's has an analogue in Schrödinger theory: Let $H=H_o+V$ be a Schrödinger Hamiltonian of a non-relativistic, finite system: H_o is e.g. the Laplacean on $L^2(\mathbb{R}^{3N})$, and V a potential depending only on the coordinates of the N particles. For many choices of V it is then easier to construct e^{-tH} rather than directly e^{itH}, because the semigroup e^{-tH} can be obtained by using path space methods and the Feynman-Kac formula.

So far, the dynamics of all models of relativistic, scalar Bose fields that have been studied is given by a Hamiltonian which is a formal generalization of H_o+V to infinitely many degrees of freedom. The transition (I.3) →(I.4) from W_n's to S_n's is related to the transformation of a hyperbolic problem to an elliptic problem + analytic continuation. We now give a rigorous definition of the S_n's.

I.4 Euclidean Description of r.q.f.t.

A rigorous version of the transition from (I.3) to (I.4) is the contents of theorems of Bargmann, Hall and Wightman and of Jost; see [SW,J]. Using (W3) (i.e. $H\geq 0$) and relativistic invariance of spec(H,P) one concludes that $W_n(x_1,\ldots,x_n)$ is the boundary value of a function $W_n(\zeta_1,\ldots,\zeta_n)$ holomorphic in a domain $\tau_n=\{\zeta_1,\ldots,\zeta_n:\mathrm{Im}(\zeta_{j+1}-\zeta_j)\in V_+\}$. This domain of holomorphy can be enlarged by means of Poincaré invariance (W2). This is the BHW theorem. Jost has then shown that locality (W4) implies that $W_n(\zeta_1,\ldots,\zeta_n)$ is holomorphic on a domain containing

$$\tau_n^{\mathrm{perm.}}=\{\zeta_1,\ldots,\zeta_n:\mathrm{Im}(\zeta_{\pi(j+1)}-\zeta_{\pi(j)})\in V_+,\pi\in\gamma_n\},$$

where γ_n are all permutations of $\{1,\ldots,n\}$, and the domain constructed in this way contains the Euclidean points

$$\mathcal{E}_n=\{\zeta_1,\ldots,\zeta_n:\zeta_j=(it_j,\vec{x}_j),(t_j,\vec{x}_j)\in\mathbb{R}^d,$$

$$\zeta_k\neq\zeta_\ell,\text{ for }k\neq\ell\} \qquad (I.5)$$

The rigorous definition of S_n is:

$$S_n(x_1,\ldots,x_n) = W_n(it_1,\vec{x}_1,\ldots,it_n,\vec{x}_n) \qquad (I.6)$$

As follows from Jost's theorem the S_n's have the attractive feature that they are <u>symmetric</u> under permutations of their arguments; (follows from holomorphy properties of W_n's and locality). The symmetry property is <u>compatible</u> with the S_n's being the moments of some measure [BY] and hence the use of path space techniques and functional integrals. (It does however <u>not</u> imply that the S_n's are moments of a measure).

The following <u>program for c.q.f.t.</u> is due to Symanzik and Nelson:

In order to construct a r.q.f. model, try to

1) directly construct $\{S_n\}_{n=o}^{\infty}$ (using path space methods and functional integrals).
2) recover $\{W_n\}$ by analytic continuation in the time variables.

This program raises the following <u>crucial question</u>: What properties of a sequence of S_n's guarantee that they are the EGF's in the sense of equation (I.6) of a unique r.q.f.t. described in terms of W_n's?

An <u>answer</u> has been given in an important analysis of Osterwalder and Schrader [OS]. We quote their result in a somewhat vague form:

<u>Theorem I.1</u> (Osterwalder-Schrader reconstruction)

<u>Hypotheses</u> (called O-S axioms)

(EO) The S_n's are tempered <u>distributions</u> the order of which in dependence of n (the nb. of points) is restricted by some <u>growth condition</u>, (see [OS])

(E1) The S_n's are <u>Euclidean invariant</u>.

(E2) Let $\mathscr{S}_<$ be the class of sequences $\underline{f}$ in $\underline{\mathscr{S}}$ such that, for all n,
$\text{supp } f_n \subseteq \{x_1,\ldots,x_n : 0 \leq t_1 < t_2 < \ldots < t_n\}$
Define $\theta\underline{f}$ by $\theta f_n\ (t_1,\vec{x}_1,\ldots,t_n,\vec{x}_n) =$
$= f_n(-t_1,\vec{x}_1,\ldots,-t_n,\vec{x}_n)$.
Then, for all $\underline{f}$ in $\underline{\mathscr{S}}_<$

$$\sum_{n,m} S_{m+n}(\theta f_m^* \otimes f_n) \geq 0.$$

Remark: This inequality "follows" easily from the heuristic expression (I.4) for the S_n's.

(E3) The S_n's are symmetric under permutations of their arguments.

Conclusion: The S_n's are the EGF's of a unique r.q.f.t. determined by W_n's satisfying (WO)-(W4), (in the sense of equation (I.6)).

Exercise 2: Find a heuristic proof of (WO)-(W4) => "(EO)", (E1)-(E3), starting from the heuristic equations (I.3), (I.4) and from (I.6). This is relatively straightforward. The converse, namely (EO)-(E3) => (WO)-(W4), is difficult. The reader should consult [OS].

I.5 Euclidean Field Theory (E.f.t.)

(or: Nelson-Symanzik Positivity)

In this section we are dealing with the question whether the S_n's are themselves v.e.v.'s of some Euclidean field, (i.e., in view of (E3), whether they are the moments of some probability measure). In general the answer is expected to be no. In important special cases it is however yes. Such a special case is the following

Example: The free field

Recall equation (I.1):

$$W_2(x,y) = (2\pi)^{-d+1} \int \frac{d\vec{p}}{2\omega(\vec{p})} e^{i\omega(\vec{p})(s-t)} e^{-i\vec{p}(\vec{y}-\vec{x})},$$
$$\text{with } x = (t,\vec{x}),\ y=(s,\vec{y}). \tag{I.7}$$

Analytic continuation of equation (I.7) to the Euclidean points $\mathcal{E}_2$ yields

$$S_2(x,y) = (2\pi)^{-d+1} \int \frac{d\vec{p}}{2\omega(\vec{p})} e^{-\omega(\vec{p})|s-t|} e^{-i\vec{p}(\vec{y}-\vec{x})}$$
$$= (2\pi)^{-d} \int dp \frac{e^{ip(x-y)}}{p^2+m^2} \tag{I.8}$$

(Compute the r.h.s. of (I.8) by means of a contour integral!) The r.h.s. of (I.8) is the kernel of the positive operator $C \equiv (-\Delta+m^2)^{-1}$, ($\Delta$ is the Laplacean on $L^2(\mathbb{R}^d)$).

From equation (I.2) and (I.8) we now get

$S_{2n+1} = 0$, $n=0,1,2,\ldots$, and

$$S_{2n}(x_1,\ldots,x_{2n}) = \sum_{\text{pairings}} S_2(x_{i_1},x_{j_1})\ldots S_2(x_{i_n},x_{j_n}) \tag{I.9}$$

and, after smearing out (I.9) with test functions,

$$S_{2n}(f_1,\ldots,f_{2n}) = \sum_{\text{pairings}} (f_{i_1},C\,f_{j_1})\ldots(f_{i_n},Cf_{j_n}). \tag{I.10}$$

In particular

$$S_{2n}(f,\ldots,f) = (2n-1)(2n-3)\ldots 3.1(f,Cf)^n \tag{I.11}$$

Next we compare equations (I.10) and (I.11) with certain Gaussian integrals:

Let $\vec{z} = (z_1,\ldots,z_M)$ and $\vec{a}_i$ be vectors in $\mathbb{R}^M$, A a positive definite matrix on $\mathbb{R}^M$, so that $C \equiv A^{-1} \geq 0$. Then

$$N^{-1}\int \prod_1^{2n} (\vec{a}_i,\vec{z})\; e^{-1/2(\vec{z},A\vec{z})} d^M z$$
$$= \sum_{\text{pairings}} (\vec{a}_{i_1},C\,\vec{a}_{j_1})\ldots(\vec{a}_{i_n},C\,\vec{a}_{j_n}), \tag{I.12}$$

(with $N = (\pi^M(\det A)^{-1})^{1/2}$ a normalization factor).

Comparing (I.10) with (I.12) we conclude: <u>The S_n's are the moments of a Gaussian measure $d\mu_o$ with mean 0 (<=> $S_1 = 0$) and covariance (operator) $C = (-\Delta+m^2)^{-1}$</u>.

In order to define $d\mu_o$ rigorously it is convenient to compute its <u>Fourier transform</u> (characteristic functional):

$$\begin{aligned}
\int d\mu_o(\phi)\; e^{i\phi(f)} &\equiv \langle e^{i\phi(f)}\rangle_o \\
&= \sum_{n=o}^{\infty} \frac{i^n}{n!} \langle \phi(f)^n\rangle_o \\
&\equiv \sum_{n=o}^{\infty} \frac{i^n}{n!} S_n(f,\ldots,f) \\
&= \sum_{n=o}^{\infty} \frac{(-1)^n}{(2n)!} S_{2n}\;(f,\ldots,f) \\
&= e^{-1/2(f,C,f)} \qquad\qquad (I.13)
\end{aligned}$$

and we have used (I.11).

Equation (I.13) combined with a general theorem of Minlos (concerning probability measures on nuclear spaces) tells us that $d\mu_o$ can be realized as a probability measure on (the σ-algebra generated by the Borel cylinder sets in) $\mathscr{S}' \equiv \mathscr{S}'_{real}(\mathbb{R}^d)$. The "coordinate function" ϕ on $\mathscr{S}'$ ($\phi(f)[T] \equiv T(f)$, for all $T \in \mathscr{S}'$) is called the <u>Euclidean field</u>, ϕ together with $\langle - \rangle_o$ the <u>free Euclidean field</u>. (The free Euclidean field is a good random field).

We now discuss the most important <u>Properties of the free Euclidean field</u>:

(A) S_n's = moments of the Gaussian measure $d\mu_o$ on $\mathscr{S}'$ with mean O and covariance $(-\Delta+m^2)^{-1}$
= functional derivatives of $\langle e^{i\phi(f)}\rangle_o$

The measure $d\mu_o$ is <u>Euclidean invariant</u>; (this is equivalent to $\langle e^{i\phi(f)}\rangle_o = \langle e^{i\phi(f_\beta)}\rangle_o$, for all Euclidean transformations β of $\mathbb{R}^d$. Here $f_\beta(x) = f(\beta^{-1}x))$. $\phi : f \in \mathscr{S} \equiv \mathscr{S}_{real}\,(\mathbb{R}^d) \to \phi(f)$ is a well defined random field.

<u>Definition:</u> A r.q.f.t. the EGF's of which are the moments of a probability measure on $\mathscr{S}'$ is called <u>Nelson-Symanzik positive</u>.

(B) Osterwalder-Schrader (or:physical) positivity:
Let $f \in \mathscr{S}(\mathbb{R}^d)$ be "supported on positive times", (i.e. supp $f \subseteq \{t \geq 0\} \equiv \{(t,\vec{x}) \in \mathbb{R}^d : t \geq 0\}$), and set $\theta f(t,\vec{x}) = f(-t,\vec{x})$. Then

$$(\overline{\theta f}, C f) = \int dt\, ds\, \overline{f(-t,\vec{x})}\, C\,(t-s,\vec{x}-\vec{y}) f(\vec{y},s)$$

(with $C(t-s,\vec{x}-\vec{y})$ given by the r.h.s. of (I.8))

$$= (2\pi)^{-d+1} \int d\vec{p} \int dt\, ds\, \overline{\tilde{f}(t,\vec{p})}\, \tilde{f}(s,\vec{p})$$

$$\times\, e^{-\omega(\vec{p})(t+s)} (2\omega(\vec{p}))^{-1}$$

$$= (2\pi)^{-d+1} \int \frac{d\vec{p}}{2\omega(\vec{p})} \left| \int dt\, \tilde{f}\,(t,\vec{p}) e^{-t\omega(\vec{p})} \right|^2$$

$$\geq 0. \qquad \text{(I.14)}$$

Hence, for $f = \sum_{j=1}^{M} c_j\, g_j$, $g_j \in \mathscr{S}$ supported on positive times,

$$C^{\theta}_{ij} \equiv (\theta g_i, C\, g_j) \qquad \text{(I.15)}$$

are the matrix elements of a positive definite matrix on $\mathbb{C}^M$.

Therefore, for arbitrary complex numbers $c_1, \ldots, c_M$,

$$\sum_{j,k} \bar{c}_j\, c_k \langle e^{i\phi(g_k - \theta g_j)} \rangle_o$$

$$= \sum_{j,k} \{\bar{c}_j\, e^{-1/2(g_j, C\, g_j)}\}\{c_k\, e^{-1/2(g_k, C\, g_k)}\}$$

$$\times\, e^{C^{\theta}_{jk}}$$

$$\geq 0. \qquad \text{(I.16)}$$

Exercise 3: How is (I.16) related to O-S positivity in the form of inequality (E2)? Is it stronger? Show that, in the case of the free field, (I.16) and (E2) are equivalent. (See [F1, Section 1] for such results).

Let Σ_+ denote the smallest σ-algebra on $\mathscr{S}'$ with the property that all functions on $\mathscr{S}'$ spanned by $\{e^{i\phi(f)} : f \in \mathscr{S}$, f is supported on positive times$\}$ are Σ_+-measurable. A Σ_+-measurable function on $\mathscr{S}'$ is said to be localized at positive times. (If F is localized at positive times then, for all ϕ and $\tilde{\phi}$ with $\phi \restriction \{t \geq 0\} =$

$= \tilde{\phi} \upharpoonright \{t \geq 0\}$, $F(\phi) = F(\tilde{\phi})$, almost surely).

We define

$$\theta F(\phi) \equiv F(\phi_\theta),$$

where

$$\phi_\theta(f) \equiv \phi(\theta f).$$

Combining now inequality (I.16) with a trivial limiting argument we obtain

Theorem I.2 (see [F1] for a more general result)
If $F \in L^2(\mathscr{S}', d\mu_0)$ is localized at positive times then

$$\langle \overline{\theta F}\, F \rangle_0 \geq 0 \tag{I.17}$$

Remark: This is the proper probabilistic version of physical positivity in the form (E2).

(C) Exponential bounds ([F1,2]):
Let $h \in \mathscr{S}_{\mathrm{real}}(\mathbb{R}^{d-1})$. Then

$$\begin{aligned} \langle e^{\phi(h \otimes \chi_{[0,t]})} \rangle_0 &= e^{1/2 (h \otimes \chi_{[0,t]}, C h \otimes \chi_{[0,t]})} \\ &\leq e^{t \cdot 1/4 \int d\vec{p}\, |h(\vec{p})|^2 \omega(\vec{p})^{-2}} \\ &= e^{t \cdot P(h)} \end{aligned} \tag{I.18}$$

where P is a continuous functional on $\mathscr{S}(\mathbb{R}^{d-1})$. In the derivation of (I.18) we have used (I.13) and (I.8). Next we want to abstract from the simple, special case of the free field.

I.6 Quantum Measures

Let $d\mu$ be some probability measure on $\mathscr{S}'$; $d\mu$ is called a quantum measure if it satisfies the hypotheses of the following

Theorem I.3 [F1]

Hypotheses:
(A) $d\mu$ is Euclidean invariant
(B) Let $\langle - \rangle$ denote expectation with respect to $d\mu$. For all $F \in L^2(\mathscr{S}', d\mu)$ localized at positive times, (i.e. $F \in L^2(\mathscr{S}', \Sigma_+, d\mu)$)

$$\langle \overline{\theta F}\, F \rangle \geq 0 \quad \text{(O-S positivity)}$$

(C) There is some $p \in [1,2)$ and a finite constant K such that, for all $f \in L^1(\mathbb{R}^d) \cap L^p(\mathbb{R}^d)$,

$$\langle e^{\phi(f)} \rangle \leq e^{K\,(||f||_1 + ||f||_p^p)}$$

Conclusion: All the moments

$$S_n(f_1,\ldots,f_n) \equiv \langle \prod_{j=1}^{n} \phi(f_j) \rangle$$

of $d\mu$ exist. The S_n's are the EGF's (in the sense of equ. (I.6)) of a unique r.q.f.t. satisfying (WO)-(W4).

Remarks: If $f = h \otimes \chi_{[o,t]}$, with $h \in L^1(\mathbb{R}^{d-1}) \cap L^p(\mathbb{R}^{d-1})$ then (C) implies

$$\langle e^{\phi(h \otimes \chi_{[o,t]})} \rangle \leq e^{K' \cdot t}$$

with $K' = K\,(||h||_1 + ||h||_p^p)$. This inequality implies quadratic form bounds of the quantum field in terms of the Hamiltonian of the theory. It is easy to check that hypothesis (C) implies the (weaker!) hypotheses (C1)-(C3) of Theorem 2.1 of ref. [F1]. (See also [DF] and Glimm-Jaffe [Ca]).

Moreover hypotheses (A)-(C) obviously imply the Osterwalder-Schrader axioms (Theorem I.1). The point is that because of the fact that (A)-(C) are stronger than these axioms some of the sophisticated methods of [OS] can be avoided in the proof of Theorem I.3 and that one obtains somewhat stronger conclusions.

Partial proof:

(i) Construction of a Hilbert space:
Let $G, F \in L^2(\mathcal{S}', \Sigma_+, d\mu)$ be localized at positive times. Then, by (B),

$$\langle F, G \rangle \equiv \langle \overline{\theta F}\, G \rangle$$

defines a positive semi-definite inner product on $L^2(\mathcal{S}', \Sigma_+, d\mu)$. Let N be its kernel. Then

$$\mathcal{H} \equiv \overline{L^2(\mathcal{S}', \Sigma_+, d\mu)/N}^{\langle\cdot,\cdot\rangle} \qquad \text{(I.19)}$$

is a separable Hilbert space.

We denote by $v(F)$ the equivalence class of $F \in L^2(\mathscr{S}', \Sigma_+, d\mu)$ mod. N and set

$$\langle v(F), v(G)\rangle \equiv \langle F, G\rangle = \langle \overline{\theta F}\; G\rangle .$$

(This defines the scalar product on $\mathscr{H}$).

(ii) Construction of a contraction semigroup:
For all $t \geq 0$, set

$$F_t(\phi) \equiv F(\phi_t), \text{ with}$$

$\phi_t(f) \equiv \phi(f_t)$, and $f_t(s,x) = f(s-t,x)$, all $f \in \mathscr{S}$.
Obviously

$$(F_t)_s = F_{t+s} \qquad \text{(I.20)}$$

If F is localized at positive times then so is F_t. For F and G localized at positive time we have

$$\langle v(F), v(G_t)\rangle = \langle v(F_t), v(G)\rangle , \qquad \text{(I.21)}$$

Using (I.20) and (I.21) we conclude that

$$U_t\; v(F) \equiv v(F_t) \qquad \text{(I.22)}$$

defines a densely defined, symmetric semigroup on $\mathscr{H}$.

<u>Exercise 4:</u> Prove (I.21) and assertion (I.22). We now show that U_t is actually a <u>contraction semigroup</u>: For all $F \in L^2(\mathscr{S}', \Sigma_+, d\mu)$

$$\begin{aligned} 0 &\leq \langle v(F),\; U_t\; v(F)\rangle \\ &\leq \langle v(F), v(F)\rangle^{1/2} \langle v(F), U_{2t}\; v(F)\rangle^{1/2} \\ &\leq \cdots\cdots \\ &\leq \langle v(F), v(F)\rangle^{\sum_{n=1}^{N} 2^{-n}} \langle v(F), U_{2^N t}\; v(F)\rangle^{2^{-N}} \\ &\leq \langle v(F), v(F)\rangle^{\sum_{n=1}^{N} 2^{-n}} (||F||_2^2)^{2^{-N}} \end{aligned} \qquad \text{(I.23)}$$

The r.h.s. of (I.23) converges to $\langle v(F), v(F)\rangle$, as $N \to \infty$. (<u>Exercise 5:</u> Prove (I.23)).
Thus, on a dense set, $0 \leq U_t \leq \mathbb{1}$, for all $t \geq 0$. Therefore

U_t extends to all of $\mathcal{H}$. Since, for $F \in L^2(\mathcal{S}', \Sigma_+, d\mu)$, $F_t \to F$, in L^2, as $t \searrow 0$,

$$\underset{t \searrow 0}{\text{s-lim}}\, U_t = \mathbb{1}.$$

Therefore $U_t = e^{-tH}$, for some positive, selfadjoint operator H.

<u>Remark:</u> So far we have just reformulated results of [OS] in a probabilistic setting. The remaining portion of the proof of this theorem is too difficult to be reproduced here; (see [F1], also [DF]). It is based in part on ideas of [N3]. The complete proof yields the following

<u>Interpretation:</u>
- The space $\mathcal{H}$ is the physical Hilbert space of a r.q.f.t. (satisfying (Wo)-(Wv); see (Wo), section I.1).
- $\Omega \equiv v(1)$ is the physical vacuum; (see (Wiii) and (Wv)).
- The operator H is the Hamiltonian; (see (Wii), (Wiii), section I.1).
- $\phi_o(h) \equiv \phi(h \otimes \delta_o)$ is a well defined quadratic form on the domain $D(H^{1/2})$ of $H^{1/2}$; (form domain of H). It is the time O-quantum field.

  $$\phi(f) = \int dt\, e^{itH}\, \phi_o(f(t,\cdot))e^{-itH}$$

  (weakly) is the quantum field; (see (Wi)).
- $$\langle e^{\phi(h \otimes \chi_{[0,t]})} \rangle = \langle \Omega, e^{-t(H-\phi_o(h))}\Omega \rangle \qquad (I.24)$$

  (Feynman-Kac formula).

<u>Remark:</u> As a consequence of Theorem I.3 and results of [GJ5], see also [N5], one has: For <u>real</u> f in $\mathcal{S}(\mathbb{R}^d)$, $\phi(f)$ is <u>essentially selfadjoint</u> on any core for H, in particular on the domain $\{\mathcal{P}(\phi)\Omega\}$ of axiom (Wv).

<u>Corollary I.4</u> [DF] Under the hypotheses of Theorem I.3 all the bounded functions of $\{\phi(f): f \in \mathcal{S}\}$ generate a local net of von Neumann algebras satisfying <u>all</u> the Haag-Kastler axioms, [HK].

<u>Remark:</u> Under the same hypotheses, another local net can be constructed from the above one that satisfies these axioms and, in addition, <u>duality</u>; see [DF]. This follows from the deep results of [BW].

Concluding remarks:

1. Selfadjointness of $\phi(f)$, for $f \in \mathscr{S}$, and Corollary I.4 do not seem to follow from the hypotheses of Theorem I.1. Nelson-Symanzik positivity (i.e. the existence of a measure) is however unimportant: Theorem I.3 and Corollary I.4 remain true without this assumption; see [DF]. For a discussion of the Haag-Kastler axioms, see [HK, GJ4].

2. One can do a lot of functional analysis on the Hilbert space $\mathscr{H}$ reconstructed from a quantum measure $d\mu$ (thanks to O-S positivity (B)). In this way important bounds on the expectations of unbounded functions F on $\mathscr{S}'$ w.r. to $d\mu$ can be obtained. See e.g. [FS] and refs. given there.

3. Additional "axiomatic" and foundational results of direct relevance to c.q.f.t. may be found in
   - an admirable analysis of Spencer and Zirilli, [Sp2]; (Bethe-Salpeter kernel, low energy unitarity; see also refs. given there).
   - [B] (sufficient conditions on $\{S_n\}_{n=0}^{\infty}$ that imply existence of isolated one particle states; the results of [B] are based on [GJS3,Sp 1,2]).
   - [EEF] (construction of time-ordered Green's functions their relation to the EGF's, p-space analyticity; these results are of course based on deep work of [Ru,BEG,OS] and many others).

II. CONSTRUCTION OF NON-TRIVIAL (NON-GAUSSIAN) QUANTUM MEASURES

(A program for the construction of non-trivial models). The basic idea of this part is to combine the structure discussed in sections I.5 and I.6,i.e. Euclidean field theory and Nelson-Symanzik positivity with Lagrangean field theory; ("marry" functional integrals with the Lagrangean formalism). This is achieved by means of a Euclidean version of the Gell-Mann-Low formula.

This formula suggests a procedure for the construction of non-trivial quantum measures, (i.e. models with $S \neq \mathbb{1}$). We start with a brief Review of the Gell-Mann-Low formula: Let ϕ be e.g. a (real), scalar Bose field, the dynamics of which is given in terms of a formal Lagrangean:

$$\mathcal{L}(\phi) = \mathcal{L}_o(\phi) - \mathcal{L}_I(\phi) \qquad \text{(II.1)}$$

where

$$"\mathcal{L}_o(\phi) = \frac{1}{2}\int \{:\dot{\phi}(\vec{x})^2: - :(\nabla\phi)(\vec{x})^2: - m^2 :\phi(\vec{x})^2:\} d\vec{x}" \qquad \text{(II.2)}$$

$$"\mathcal{L}_I(\phi) \stackrel{e.g}{=} \lambda \int :\phi(\vec{x})^4: d\vec{x} + \underline{\text{local}} \text{ counterterms}"$$
$$\equiv \int \mathcal{L}_I(\phi(\vec{x})) d\vec{x}. \qquad \text{(II.3)}$$

Let $\tau_n(x_1,\ldots,x_n)$ denote the time-ordered Green's functions (chronological functions). Formally, (e.g. in the sense of perturbation theory), the τ_n's are given by

$$\tau_n(x_1,\ldots,x_n)$$
$$= \lim_{\Lambda \nearrow \mathbb{R}^d} \frac{\left\langle T\left(\prod_1^n \phi(x_i) e^{i\int_\Lambda dx \ \mathcal{L}_I(\phi(x))}\right)\right\rangle_o}{\left\langle T\left(e^{i\int_\Lambda dx \ \mathcal{L}_I(\phi(x))}\right)\right\rangle_o} \qquad \text{(II.4)}$$

where T stands for time-ordering and $\langle - \rangle_o$ is the v.e.v. of the free, scalar field. We now continue the τ_n's formally in the time variables to the Euclidean points:

$$\tau_n(t_1,\vec{x}_1,\ldots,t_n,\vec{x}_n) \to \tau_n(e^{i\theta}t_1,\vec{x}_1,\ldots,e^{i\theta}t_n,\vec{x}_n),$$
$$\theta \in [0,\tfrac{\pi}{2}]. \qquad \text{(II.5)}$$

As θ tends to $\frac{\pi}{2}$ we obtain the <u>EGF's:</u>

$$\tau_n(it_1,x_1,\ldots,it_n,x_n) = S_n(x_1,\ldots,x_n). \qquad \text{(II.6)}$$

Given a r.q.f.t. satisfying (WO)-(W4), it is an interesting problem (important for a proof of $S \neq \mathbb{1}$) to construct τ_n's satisfying the usual properties of chronological functions and (II.6). For theories satisfying a special version of the Osterwalder-Schrader axioms (EO)-(E3) this problem has been solved in [EEF]; (moreover it has been shown there that for some two dim. models the perturbation series of the r.h.s. of (II.4) in $\mathcal{L}_I$ is asymptotic to the exact τ_n's; see also [D2]).

Let us now do a formal analytic continuation of (II.4) to the Euclidean points, as prescribed in (II.4), (II.5), we then obtain

$$S_n(x_1,\ldots) = \tau_n(it_1,\vec{x}_1,\ldots)$$

$$= \lim_{\Lambda \nearrow \mathbb{R}^d} \frac{\int_{\mathcal{S}'} d\mu_o(\phi) \prod_1^n \phi(x_i) e^{-U_\Lambda(\phi)}}{\int_{\mathcal{S}'} d\mu_o(\phi) e^{-U_\Lambda(\phi)}} \qquad \text{(II.7)}$$

where $U_\Lambda(\phi) \equiv \int_\Lambda dx\ \mathcal{L}_I(\phi(x))$ is the "Euclidean action" This is the Euclidean Gell-Mann-Low formula.

Exercise 6: Give a formal proof (e.g. in the sense of perturbation theory) of (II.7), starting from (II.3)-(II.5). (See also [Sy 1,2, S1, N2]). A formal expression for $d\mu_o(\phi)$ is suggestive:

$$\text{"}d\mu_o(\phi) = N^{-1} e^{-1/2\int\{:(\nabla\phi)(x)^2:+m^2:\phi(x)^2:\}dx} \times \prod_{x\in\mathbb{R}^d} \mathcal{D}\phi(x)\text{"} \qquad \text{(II.8)}$$

The basic problem with the Euclidean G-L formula:

In order to get non-trivial interactions, i.e. models with $S\neq\mathbb{1}$, the degree of $\mathcal{L}_I$ (in our example=4) must be ≥ 4, i.e. $\mathcal{L}_I(\phi(x))$ must be a sum of terms of the form $\phi(x)^n$; or of

$$\text{"}:\phi(x)^n: \equiv \sum_{m=o}^{[n/2]} \frac{n!}{m!(n-2m)!} \phi(x)^{n-2m} \left(-\tfrac{1}{2}\langle\phi(x)^2\rangle_o\right)^m \text{"}$$

(Wick order), (II.9)

with $n_{\text{máx.}} \geq 4$.

However the support of $d\mu_o$ ($\subset \mathcal{S}'$) can be shown to consist of distributions that are almost surely not functions. Hence, for $\phi \in \operatorname{supp} d\mu_o$, $\phi(x)^4$ and even the more regular version $:\phi(x)^4:$ are in general not well defined random fields, (i.e. $\int dx\ \phi(x)^4 f(x)$, $f \in \mathcal{S}$, is not a random variable). Therefore, as it stands, the Euclidean G-L formula (II.7) may be meaningless. This is the famous

Problem of ultraviolet divergences.

Fact (proven in part III): in d=2 space-time dimensions

$$\int dx :\phi(x)^n: f(x), \quad n=1,2,3,\ldots,$$

$$f \overset{\text{e.g.}}{\in} \mathcal{S}(\mathbb{R}^2), \text{ is a well defined} \tag{II.10}$$

random variable in $\bigcap_{p<\infty} L^p(\mathcal{S}', d\mu_0)$.

Assume for a moment that

(i) there are no ultraviolet divergences, i.e. $U_\Lambda(\phi)$ is a well defined random variable; (e.g. $d \overset{\Lambda}{=} 2$; see (II.10)).

(ii) $e^{-U_\Lambda(\phi)}$ is $d\mu_0$-integrable.

Then

$$d\mu_{U_\Lambda}(\phi) \equiv \left(\int_{\mathcal{S}'} e^{-U_\Lambda(\phi)} d\mu_0(\phi)\right)^{-1} e^{-U_\Lambda(\phi)} d\mu_0(\phi) \tag{II.11}$$

is a well defined probability measure on $\mathcal{S}'$, (called "cutoff interacting measure").

Lemma II.1: Suppose that Λ is invariant under reflections at the hyperplane $\{t=0\}$, i.e.

$$\Lambda = \Lambda_+ \cup \theta\Lambda_+, \text{ where } \Lambda_+ \subset \{t \geq 0\},$$

and

$$e^{-U_\Lambda} = e^{-(U_{\Lambda+} + U_{\theta\Lambda+})} = e^{-U_{\Lambda+}} e^{-U_{\theta\Lambda+}}, \tag{II.12}$$

with $e^{-U_{\Lambda+}}$ localized at positive times.
Then $d\mu_{U_\Lambda}$ satisfies O-S positivity:

$$\int d\mu_{U_\Lambda}(\phi)\ \overline{\theta F(\phi)}\ F(\phi) \equiv \langle \overline{\theta F(\phi)}\ F(\phi) \rangle_{U_\Lambda} \geq 0 \tag{II.13}$$

for all F's localized at positive times.

Exercise 7: Starting with definitions (II.2),(II.3) for d=2, and (II.7), use (II.10) to prove (II.12), the equation: $e^{-U_{\theta\Lambda+}} = \Theta(e^{-U_{\Lambda+}})$, and Lemma II.1.

Lemma II.2: Let $\{\Lambda_n\}_{n=o}^{\infty}$ denote an arbitrary sequence of rectangles converging to $\mathbb{R}^d$, by inclusion, ($\Lambda_n \subset \Lambda_{n+1}$, for all n), and assume that

$$d\mu_U = \lim_{n\to\infty} d\mu_{U_{\Lambda_n}} \text{ exists}$$

and is independent of the choice of $\{\Lambda_n\}_{n=o}^{\infty}$; (limit e.g. in the sense of convergence of characteristic functionals). Then

(a) $d\mu_U$ is Euclidean invariant, and

(b) $d\mu_U$ satisfies O-S positivity, (i.e. hypotheses (A) and (B) of Theorem I.3 hold).

The proof is almost trivial. To show that $d\mu_U$ is a quantum measure it now suffices to establish the exponential bounds of hypothesis (C) of Theorem I.3. General conditions under which these bounds are valid are discussed in [F1, FS]. As they are somewhat technical we omit a discussion. From (II.11) and Lemmata II.1 and II.2 we now derive the following

Program for the construction of non-trivial models:

1) Define $d\mu_{U_\Lambda}$ rigorously as a probability measure on $\mathcal{S}'$ for all bounded rectangles Λ, (and prove (II.12), which in d=3 or more dimensions is in general not automatic; "Renormalization of UV divergences").
2) Prove that $\lim_{\Lambda\nearrow\mathbb{R}^d} d\mu_{U_\Lambda}$ exists, in the sense of Lemma II.2; ("Thermodynamic limit and infrared divergences").
3) Verify the exponential bounds of hypothesis (C) of Theorem I.3; ("Stability under local perturbations of the dynamics"; see (C) and (I.24); also [F1,2]).

Once this program is carried out in a given model one is usually interested in:

(a) Connection of constructed r.q.f.t. with perturbation theory (see [D1,2, EMS, MS, FSe, EEF] for results) and the renormalization group.
(b) Particle structure: one particle states, bound states and resonances. (As mentioned, sufficient

conditions on a sequence $\{S_n\}_{n=o}^{\infty}$ of EGF's that imply the existence of an isolated one particle shell in spec (H,P) have been derived in [B]. They are based on [GJS3,Sp1,2]. For results concerning (b) in the context of models see [GJS3, Sp1,2, FSe, DE]. These results are very technical. They are based on expansion methods developped in [GJS1,2,3,4]. See also [GRS4,GJ7]).

(c) Is $S \neq \mathbb{1}$; (see [EEF,OSé]; and [Sp1,2] for an analysis of low energy unitarity).

(d) Structure of the physical vacuum; (vacuum degeneracy, broken symmetries, etc.; see [GJS4, FSS]; also [F3]).

(e) Non-trivial super-selection rules (e.g. "quantum-solitons"; see [Co] for a review, [F3,4, St] for some rigorous (general) results about quantum solitons, [Lü] for a beautiful special result, and [DHR,R] for an axiomatic analysis of super-selection sectors).

(f) Critical theories; (investigation of the critical point, O-mass theories - see e.g. [GJ6] for some rigorous results – scaling invariance, infrared divergences; see [Ca] for a review).

III. THE THREE-STEP PROGRAM AT WORK: THE $\lambda\phi^4$ THEORY IN TWO DIMENSIONS

The $\lambda\phi_2^4$ theory ($\lambda\phi^4$ in two space-time dimensions) is not very difficult to construct and analyze compared to the Y_2- (Yukawa interaction in two dimensions, [Y_2]) or the $\lambda\phi_3^4$ theories, [ϕ_3^4].

Nevertheless the mathematical techniques developped in the study of the $\lambda\phi_2^4$ theory since 1971 (see also [G,N1]) have been essential even for the analysis of $\lambda\phi_3^4$ and, (if supplemented with some amount of theory of renormalized Fredholm determinants, [Y_2]) the one of Y_2.

We therefore hope that the subsequent rather elementary and incomplete exposition of $\lambda\phi_2^4$ communicates to the reader some of the flavour of the mathematical tools of c.q.f.t.. (Warning: Many (if not most) important technical aspects of c.q.f.t. are not treated in the following. Moreover the tools are presently in a period of rapid development. Finally an exposition of c.q.f.t. without a chapter on gauge theories may soon be felt to be incomplete; see e.g. the authors contribution to.

III.1 To Start with Something very Easy: $\lambda\phi^4{}_o$

This theory is no theory; it is just a probability measure on the real line. The free measure is

$$d\mu_o(\Phi) = N_o^{-1} e^{-\frac{m^2}{2}\phi^2} d\phi\,,\quad \phi \in \mathbb{R}$$

where $N_o = \sqrt{\frac{2\pi}{m^2}}$

The Euclidean action is

$$U(\phi) = \lambda{:}\phi^4{:} \equiv \lambda\phi^4 - 6\lambda\phi^2 \langle\phi^2\rangle_o + 3\lambda \langle\phi^2\rangle_o^2$$

$$= \lambda\phi^4 - \frac{6\lambda}{m^2}\phi^2 + \frac{3\lambda}{m^4}\,.$$

Hence $U(\phi) \geq -\frac{6\lambda}{m^2}$ with minima at $\phi = \pm\sqrt{\frac{3}{m^2}}$

The interacting measure is:

$$d\mu_U(\phi) = N^{-1} e^{-\lambda:\phi^4:} d\mu_o(\phi)\,,$$

$$N = \int_{\mathbb{R}} e^{-\lambda:\phi^4:} d\mu_o(\phi) \qquad \text{(III.1)}$$

There is no cutoff to be removed, (and hypotheses (A)-(C) of Theorem I.3 become trivial or void). Nevertheless the following simple features of $\lambda\phi_o^4$ should be noticed:

(i) The Euclidean action $U(\Phi)$ is "almost positive"; $e^{-U(\phi)}$ is $d\mu_o$-integrable.

(ii) Perturbation theory in λ diverges: Consider e.g. N(the "partition function"). If one expands $e^{-\lambda:\phi^+:}$ in (III.1) in a power series one obtains

$$N = \sum_{n=o}^{\infty} \frac{(-\lambda)^n}{n!} \int ({:}\phi^4{:})^n \, d\mu_o(\phi)\,, \qquad \text{(III.2)}$$

and an easy calculation (or an estimate) shows that

$$\int ({:}\phi^4{:})^n d\mu_o(\phi) = O((n!)^2) \qquad \text{(III.3)}$$

Hence the series on the r.h.s. of (III.2) diverges! (This situation does not change if we consider the

the moments $\int \phi^{2m} \, d\mu_U(\phi))$.

(iii) However the perturbation series of e.g.

$N, \int \phi^{2m} d\mu_U(\phi), \; m=1,2,\ldots$

is Borel-summable. This means that all the moments of $d\mu_U$ (the "EGF"s) are uniquely determined by their perturbation series.

(iv) Correlation inequalities and Lee-Yang theorem;

$$\int \phi^n \, d\mu_U(\phi) \geq 0 \tag{III.4}$$

("first Griffiths inequality"; the proofs of the simplest correlation inequalities, such as the first and second Griffiths inequality, are never much more difficult than the one of (III.4)). Finally

$$\int e^{z\phi} \, d\mu_U(\phi)$$

has its zeroes (in z) on the imaginary axis; (Lee-Yang property; this is the only statement among (i)-(iv) that is not completely trivial; see [Nw, S2]).

We have formulated these features of $\lambda\phi_0^4$ because they are typical for $\lambda\phi_1^4$, $\lambda\phi_2^4$ and to a large extent for $\lambda\phi_3^4$.

Exercise 8: Complete the three-step program for the $\lambda\phi_1^4$ theory. (This is the anharmonic oscillator with Hamiltonian $H = H_o + \lambda{:}q^4{:}$, with $H_o = \frac{1}{2}(-\frac{d^2}{dq^2} + m^2q^2)$. Here the (free) Gaussian measure $d\mu_o(\phi)$ is the path space measure associated with the transition function e^{-tH_o}; its covariance is given by the kernel $\frac{1}{2m} e^{-m|t-s|}$; $\{\phi = q(t)\}$ are the paths. Show that perturbation theory diverges).

III.2 The $\lambda\phi_2^4$ Theory: Stability

Definition: We let $\eta(k) \equiv \eta(|k|)$ be a C^∞-function on the plane $\mathbb{R}^2$ with $\eta(k) = 1$, for $0 \leq |k| \leq 1/2$, $\eta(k) = 0$, for $|k| \geq 1$, and $0 \leq \eta \leq 1$ (III.5)

$$\text{Let} \quad h_\kappa(x) = \frac{1}{(2\pi)^2} \int \eta\left(\frac{k}{\kappa}\right) e^{ikx} \, d^2k, \quad \text{with } 0<\kappa<\infty \tag{III.6}$$

We introduce an ultraviolet cutoff Euclidean field

$$\phi_\kappa(x) \equiv (h_\kappa * \phi)(x). \tag{III.7}$$

Note that $\phi_\kappa(x)$ is a function, and

$$<\phi_\kappa(x)^2>_o = \int \frac{d^2k\ \eta(k/\kappa)^2}{k^2+m^2} \leq O\ (\log\ \kappa) \tag{III.8}$$

Therefore $\phi_\kappa(x)^n$, and

$$:\phi_\kappa(x)^n: = \sum_{m=o}^{[n/2]} \frac{n!}{m!(n-2m)!}\ \phi_\kappa(x)^{n-2m}(-\tfrac{1}{2}<\phi_\kappa(x)^2>_o)^m \tag{III.9}$$

(Wick ordering of $\phi_\kappa(x)^n$!)

are well defined random variables.

The Euclidean action with ultraviolet cutoff κ and space-time cutoff Λ is given by

$$U_{\kappa,\Lambda}\ (\phi) = \lambda \int_\Lambda d^2x\ :\ \phi_\kappa(x)^4: \tag{III.10}$$

From (III.8) and (III.9) we obtain

$$:\phi_\kappa(x)^4: \geq -O((\log\kappa)^2), \text{ and}$$
$$U_{\kappa,\Lambda}\ (\phi) \geq -O((\log\kappa)^2)|\Lambda|, \tag{III.11}$$

with $|\Lambda|$ the volume of Λ.

Lemma III.1:

(a) For all $\kappa' \geq \kappa > 0$

$$||U_{\kappa,\Lambda}(\phi) - U_{\kappa',\Lambda}(\phi)||_2^2$$
$$\equiv \int_{\mathscr{S}'} d\mu_o(\phi)|U_{\kappa,\Lambda}\ (\phi)-U_{\kappa',\Lambda}\ (\phi)|^2$$  
$$= \kappa \text{ [diagram] } \kappa \ - \ \kappa' \text{ [diagram] } \kappa' \tag{III.12}$$
$$\leq O(\kappa^{-\varepsilon}), \text{ for some } \varepsilon>0.$$

(b) $U_\Lambda(\phi) \equiv \underset{\kappa\to\infty}{\text{s-lim}}\, U_{\kappa,\Lambda}(\phi)$

exists in $L^2(\mathcal{S}', d\mu_o)$. If $\Lambda \subseteq \{t\geq 0\}$ then $U_\Lambda(\phi)$ is <u>localized at positive times.</u>

<u>Proof:</u> (a) is a standard exercise in estimating (Euclidean region) Feynman diagrams; (depicted in (III.12)). The first part of (b) follows of course from (a), the second part follows by choosing ultraviolet cutoffs only in spatial directions and showing that the same limit is obtained, as they are removed.

Q.E.D.

The random variable $U_\Lambda(\phi)$ is the space-time cutoff Euclidean action. Part (b) and $d\mu_o$-integrability of $e^{-U_\Lambda(\phi)}$ will imply O-S positivity, (hyp. (B), Theorem I.3), of $d\mu_{U_\Lambda}(\phi)$, provided Λ is invariant under reflections at $\{t=0\}$; see Exercise 7.

Thus the basic problem to study is $d\mu_o$-integrability of e^{-U_Λ}. We seem to have some badluck: In contradistinction to the situation in $\lambda\phi^4_o$ and $\lambda\phi^4_1$, U_Λ is <u>not</u> bounded below. Nevertheless we shall show that e^{-U_Λ} <u>is</u> $d\mu_o$-integrable.

An analogy suggests why:
Consider a non-relativistic particle in the Coulomb potential $V(\vec{x}) = -\frac{e^2}{|\vec{x}|}$; V is not bounded below, yet $H \equiv -\Delta+V(\vec{x})$ <u>is.</u> The uncertainty principle tells us why: If $\langle\psi,V\psi\rangle$ is very small then $-\langle\psi,\Delta\psi\rangle$ must be very large. If one studies e^{-tH} by means of path space techniques one arrives at a problem that is analogous to the $d\mu_o$-integrability of e^{-U_Λ}.

<u>Theorem III.2:</u>

(a) e^{-U_Λ} is $d\mu_o$-integrable; [N1,G].

(b) $1\leq\int_{\mathcal{S}'} d\mu_o(\phi)e^{-U_\Lambda(\phi)} \leq e^{O(|\Lambda|)}$; [GJ8,DG,N2,GRS5]

Remark: Part (a) settles step 1) of our three-step program. Part (b) sharpens (a) and, as it turns out, settles some portion of step 2). The lower bound in (b) follows from Jensen's inequality:

$$\langle e^{-U_\Lambda}\rangle_o \geq e^{-\langle U_\Lambda\rangle_o} = 1.$$

The proof of (a) is a Euclidean reformulation of Nelson's original argument [N1]; (b) is based on work of [G,GJ8, DG].

The basic idea of the proof [N1]: The μ_o-measure of a region in $\mathscr{S}'$ on which U_Λ is very negative must be very small! This will give (a). As first shown in [GJ8,DG], the fact that the upper bound in (b) is of the form $e^{O(|\Lambda|)}$ follows from the rapid decrease of the kernel of the covariance C of $d\mu_o$:

$$C(x-y) = (2\pi)^{-2}\int \frac{e^{ip(x-y)}}{p^2+m^2}\, d^2p$$

$$\simeq e^{-m|x-y|}, \text{ as } |x-y| \to \infty$$

From this and definitions (III.5) and (III.6) it then follows that

$$\langle \Phi_\kappa(x)\Phi_{\kappa'}(y)\rangle_o \text{ decreases in } |x-y| \text{ faster than any power of } |x-y|^{-1}! \tag{III.13}$$

(This leads to the intuition that $d\mu_o$ "almost decouples" distant space-time regions, or, "$\Phi_\kappa(x)$ and $\Phi_{\kappa'}(y)$ are almost independent random variables, for $|x-y|$ large". Hence (b) should essentially follow from (a). See e.g. [N2]).

Proof:

1) Cover $\mathbb{R}^2$ with a grid of unit squares $\{\Delta_j\}_{j\in\mathbb{Z}^2}$ centered at the sites of $\mathbb{Z}^2$ with faces parallel to the coordinate axes. Assume for simplicity that $\Lambda = \bigcup_{j\in\tilde{\Lambda}} \Delta_j$, $\tilde{\Lambda}$ a bounded subset of $\mathbb{Z}^2$.

By (III.10) and Lemma III.1, (b)

$$U_\Lambda = \sum_{j\in\tilde{\Lambda}} U_{\Delta_j}, \text{ i.e. } e^{-U_\Lambda} = \prod_{j\in\tilde{\Lambda}} e^{-U_{\Delta_j}} \tag{III.14}$$

Let $I_n = [-n,-n+1)$, and

$$I_o = (0,\infty).$$

Moreover $\underline{n} = \{n_j\}_{j\in\tilde{\Lambda}}, |\underline{n}| = \sum_{j\in\tilde{\Lambda}} n_j$

We define regions $M_{\underline{n}} \subset \mathcal{S}'$ on which a lower bound for U_Λ is known. (We shall then derive estimates on the μ_o-measure of $M_{\underline{n}}$).

$$M_{\underline{n}} = \{\Phi : U_{\Delta_j}(\Phi) \in I_{n_j},\ j \in \tilde{\Lambda}\}$$

$$= \bigcap_{j\in\tilde{\Lambda}} \{\Phi : U_{\Delta_j}(\Phi) \in I_{n_j}\} \qquad \text{(III.15)}$$

By Lemma III.1, (b) $\{U_{\Delta_j}\}_{j\in\tilde{\Lambda}}$ are measurable functions, therefore $M_{\underline{n}}$ is a measurable set, for all $\underline{n}$. Clearly

$$\bigcup_{\underline{n}} M_{\underline{n}} = \mathcal{S}' \qquad \text{(III.16)}$$

By (III.14),(III.15) and (III.16)

$$\int_{\mathcal{S}'} d\mu_o(\Phi)\, e^{-U_\Lambda(\Phi)} \leq \sum_{\underline{n}} e^{|\underline{n}|} \mu_o(M_{\underline{n}}) \qquad \text{(III.17)}$$

2) Estimating $\mu_o(M_{\underline{n}})$:

Define κ_{n_j} such that

$$U_{\kappa_{n_j},\Delta_j} \geq -n_j + 2;\ n_j = 1,2,\ldots \qquad \text{(III.18)}$$

Using (III.11) we see that, with

$$\kappa_{n_j} = c_1\, e^{c_2\sqrt{n_j}}, \qquad \text{(III.19)}$$

inequality (III.18) holds; (in the following c_k denotes a finite, positive constant).

Combining (III.15) and (III.18) we see that

$M_{\underline{n}}$ is contained in the set

$$\tilde{M}_{\underline{n}} \equiv \bigcap_{\substack{j\in\tilde{\Lambda}\\ n_j\geq 1}} \{\Phi : U_{\kappa_{n_j},\Delta_j}(\Phi) - U_{\Delta_j}(\Phi) \geq 1\} \tag{III.20}$$

Thus

$$\begin{aligned}\mu_o(M_{\underline{n}}) &\leq \mu_o(\tilde{M}_{\underline{n}}) \\ &\leq \int_{\tilde{M}_n} d\mu_o(\Phi) \prod_{j\in\tilde{\Lambda}} \left| U_{\kappa_{n_j},\Delta_j}(\Phi) - U_{\Delta_j}(\Phi)\right|^{p_{n_j}} \\ &\leq \int_{\mathcal{S}'} d\mu_o(\Phi) \prod_{j\in\tilde{\Lambda}} \left| U_{\kappa_{n_j},\Delta_j}(\Phi) - U_{\Delta_j}(\Phi)\right|^{p_{n_j}} \\ &\equiv \left\langle \prod_{j\in\tilde{\Lambda}} \left| U_{\kappa_{n_j},\Delta_j} - U_{\Delta_j}\right|^{p_{n_j}} \right\rangle_o,\end{aligned} \tag{III.21}$$

where $p_{n_j} = 0$, for $n_j=0$, and $p_{n_j} \geq 1$ arbitary, for $n_j\geq 1$. In particular we may choose p_{n_j} to be <u>even</u>, for all $j\in\tilde{\Lambda}$.

In this case, estimating the r.h.s. of (III.21) amounts to computing a Gaussian integral of some Wick-polynomial in the field Φ, i.e. a large (Euclidean region) <u>Feynman diagram</u>. <u>Even though such an estimate is completely elementary, it is not very easy to obtain</u>, (a typical feature of many techniques in c.q.f.t.!). A good estimate is the following

<u>Lemma III.3:</u> [DG,EMS]

$$\begin{aligned}&\left\langle \prod_{j\in\tilde{\Lambda}} \left| U_{\kappa_{n_j},\Delta_j} - U_j\right|^{p_{n_j}} \right\rangle_o \\ &\leq \prod_{j\in\tilde{\Lambda}} \sqrt{(4p_{n_j})!}\,(c_3\,\kappa_{n_j})^{-p_{n_j}\cdot\varepsilon},\end{aligned}$$

with $\varepsilon>0$ as in Lemma III.1, (a).

3) <u>Completing the proof of Theorem III.2:</u>
(assuming Lemma III.3)

$$\int_{\mathcal{S}'} d\mu_o(\Phi) e^{-U_\Lambda(\Phi)} \leq \sum_{\underline{n}} e^{|\underline{n}|} \mu_o(M_{\underline{n}})$$

$$\leq \sum_{\underline{n}} \prod_{j\in\tilde{\Lambda}} e^{n_j} \mu_o(\tilde{M}_{\underline{n}})$$

$$\leq \sum_{\underline{n}} \prod_{j\in\tilde{\Lambda}} e^{n_j} \sqrt{(4p_{n_j})!}\, (c_3\kappa_{n_j})^{-p_{n_j}\cdot\varepsilon}$$

$$= \prod_{j\in\tilde{\Lambda}} \Big[\sum_{n_j=o}^{\infty} e^{n_j} \sqrt{(4p_{n_j})!}\, (c_3\kappa_{n_j})^{-p_{n_j}\cdot\varepsilon} \Big] \tag{III.22}$$

where we have used (III.21) and Lemma III.3. Recall that $\kappa_{n_j} = c_1\, e^{c_2\sqrt{n_j}}$. Next, choose

$$p_o = 0,\ p_n \overset{\text{e.g.}}{=} 2n \tag{III.23}$$

Then

$$e^{n_j} \sqrt{(4p_{n_j})!}\, (c_3\ \kappa_{n_j})^{-p_{n_j}\cdot\varepsilon} \leq$$

$$\leq c_4\ e^{-c_5\ n_j^{3/2} + 4n_j \log (8n_j) + c_6\ n_j}$$

$$\leq c_7\ e^{-c_8\ n_j^{3/2}} \tag{III.24}$$

Hence $\sum_{n_j=o}^{\infty} e^{n_j} \sqrt{(4p_{n_j})!}\, (c_3\kappa_{n_j})^{-p_{n_j}\cdot\varepsilon} \leq c_9$, (indep. of j!)

so, using (III.22),

$$\int_{\mathcal{S}'} d\mu_o(\Phi)\ e^{-U_\Lambda(\Phi)} \leq \prod_{j\in\tilde{\Lambda}} c_9 = c_9^{|\tilde{\Lambda}|}$$

$$= e^{O(|\Lambda|)}.$$

This completes the proof of Theorem III.2.

Remarks: The basic idea of the proof, namely (III.15), (III.17), (III.21), is due to Nelson [N1]. In [N1] and [G] the localization in space-time- see (III.14),

(III.20), (III.21) - leading to the stability estimate ("linear lower bound") $e^{O(|\Lambda|)}$ was not yet introduced. This technique was invented in [GJ8] and, in a form essentially identical to the one presented here (with slightly worse combinatorial estimates) in [DG]. Other proofs of Theorem III.2, (b) were given in [N2] (combines the original argument of [N1] with "hypercontractivity" and Euclidean invariance of $d\mu_o$), [GRS5] (combines [N1,G] with the use of "Neumann boundary conditions" in the covariance of $d\mu_o$). See also [GRS3,GJS2, S1].

Note that the proof we have presented and the arguments to follow exhibit some typical aspects of <u>expansion methods</u> as developped in [GJ3,GJS2, EMS,MS,...].

4) <u>Comments on the proof of Lemma III.3:</u>

<u>Exercise 9</u>: Prove Lemma III.3 for the case of $\tilde{\Lambda}=\{j\equiv(o,o)\}$. (This yields Theorem III.2, (a)). To prove Lemma III.3 we must study the Gaussian integral

$$I \equiv \left\langle \prod_{j\in\tilde{\Lambda}} \left| U_{\kappa_{n_j},\Delta_j} - U_{\Delta_j} \right|^{p_{n_j}} \right\rangle_o, \tag{III.25}$$

(with p_{n_j}'s as in (III.23). The integrand is a Wick polynomial in Φ). <u>Wick's theorem</u> tells us that I is a sum over all possible contraction schemes forming vacuum diagrams $\mathcal{G}$; (one contraction replaces two fields in the integrand by their $d\mu_o$-integral, i.e. their free, Euclidean v.e.v.; see [DG]).

Let G denote the numerical value (number) associated with the contraction scheme (vacuum diagram) $\mathcal{G}$. Then

$$I = \sum_{\mathcal{G}} G \quad \text{(Wick's theorem)}. \tag{III.26}$$

This sum has very many terms, and a direct estimate is difficult. Glimm and Jaffe [GJ3] have invented the very efficient method of "<u>combinatoric factors</u>" to simplify such estimates:
Let $\{c(\mathcal{G})\}$ be such that

$$\sum_{\mathcal{G}} c(\mathcal{G})^{-1} \leq 1. \tag{III.27}$$

$$\text{Then } I \le \sum_{\mathcal{G}} c(\mathcal{G})^{-1}(c(\mathcal{G})|G|)$$
$$\le \sup_{\mathcal{G}} c(\mathcal{G})|G|; \qquad \text{(III.28)}$$

(more generally: if $\{\mathcal{G}\} = \{\mathcal{G}_{\underline{n}} : \underline{n} \in J \subseteq \mathbb{Z}^m\}$, $\underline{n} \equiv (n_1,\dots,n_m)$ and $\sum_{n_\ell} c_{n_\ell}^{-1} \le 1$, for all $\ell=1,\dots,m$, then

$$I = \sum_{\underline{n}\in J} G_{\underline{n}} \le \sup_{\underline{n}\in J} (c_{n_1}\dots c_{n_m}|G_{\underline{n}}|)).$$

In principle, every choice for $c(\mathcal{G})$ compatible with (III.27) is possible. However, in order to get a good estimate, one must do a <u>clever</u> choice.

Let $M \equiv |\tilde{\Lambda}|$, $\Delta^{(1)},\Delta^{(2)},\dots,\Delta^{(M)}$ the cubes $\{\Delta_j\}_{j\in\tilde{\Lambda}}$ ordered such that

$$n_1 \ge n_2 \ge \dots \ge n_m, \text{ where}$$
$$n_\ell = 4p_{n_j} \text{ if } \Delta^{(\ell)} = \Delta_j; \text{ (see [MS]).} \qquad \text{(III.29)}$$

Wick's theorem tells us to contract all $n_1+\dots+n_M = 4(p_{n_1}+\dots+p_{n_{|\tilde{\Lambda}|}})$ fields in the integrand of I (r.h.s. of (III.25)) in all possible ways among each other, (<u>leaving out</u> contractions of fields belonging to the same "vertex" $U_{\kappa_{n_j},\Delta_j} - U_{\Delta_j}$, because of Wick ordering), and replacing each contracted pair of fields by their Gaussian expectation ($d\mu_0$-integral). In this way one arrives at (III.26) The combinatorial aspects of this procedure (and a <u>suitable choice</u> for $\{c(\mathcal{G})\}$!) can be understood by studiing the following problem:

We are given M squares $\Delta^{(1)},\dots,\Delta^{(M)}$ (belonging to the covering $\{\Delta_j\}_{j\in\mathbb{Z}^2}$ introduced in 1)). We put n_ℓ <u>pucks</u> onto $\Delta^{(\ell)}$. We are asked to estimate in how many ways one can form pairs of pucks (<u>dumbells</u>) by joining <u>each</u> puck in each square to <u>one other</u> puck in the same or another square by means of a wire; ($P\equiv n_1+\dots+n_M$ is assumed to be <u>even</u>). Of course we all know what the number N of ways of forming such dumbells is:

$$N = (P-1)(P-3)(P-5)\ldots 3\cdot 1$$

However, an <u>estimate</u> (upper bound) on N (more suitable for our puposes) can be obtained as follows:

Pick a puck in $\Delta^{(1)}$. It has to be joined to some puck in $\Delta^{(\ell)}$, $\ell=1,\ldots,M$. Let us <u>first choose</u> $\Delta^{(\ell)}$. Of course there are M choices. But let us instead introduce a <u>combinatoric factor</u> for the choice of $\Delta^{(\ell)}$:

Let $d(\Delta,\Delta') \equiv$ const. $[\text{dist.}(\Delta,\Delta')+1]$. By a proper choice of the constant we achieve that

$$\sum_{\Delta^{(\ell)}\in\{\Delta_j\}_{j\in\tilde{\Lambda}}} d(\Delta^{(k)},\Delta^{(\ell)})^{-3}$$

$$\leq \sum_{\Delta'\in\{\Delta_j\}_{j\in\mathbb{Z}^2}} d(\Delta^{(k)},\Delta')^{-3} \leq 1. \qquad \text{(III.30)}$$

Comparing (III.30) with (III.27) we figure that we may use

$$d(\Delta^{(k)},\Delta^{(\ell)})^3 \qquad \text{(III.31)}$$

as a <u>combinatoric factor</u> for the choice of $\Delta^{(\ell)}$, given a puck in $\Delta^{(k)}$ to be connected to some puck in $\Delta^{(\ell)}$.

We call $d(\Delta^{(k)},\Delta^{(\ell)})$ the <u>length of the dumbell</u> obtained by joining a puck in $\Delta^{(k)}$ to one in $\Delta^{(\ell)}$. Next let k=1, and let some puck in $\Delta^{(1)}$ be given. Having chosen $\Delta^{(\ell)}$ we have n_ℓ possibilities to choose a puck in $\Delta^{(\ell)}$. Since, by (III.29), $n_1 \geq n_\ell$, we conclude that there are at most

$$\sqrt{n_1}\;\sqrt{n_\ell} \qquad \text{(III.32)}$$

choices of a puck in $\Delta^{(\ell)}$. We attribute the factor $\sqrt{n_1}$ to the given puck in $\Delta^{(1)}$ and $\sqrt{n_\ell}$ to the puck chosen in $\Delta^{(\ell)}$. Next we proceed in the same manner with the second puck in $\Delta^{(1)}$, (using (III.31) and (III.32) to estimate all possible choices), then with the third one etc., until all pucks in $\Delta^{(1)}$ have been joined to other pucks. Then we finish with joining the remaining pucks in $\Delta^{(2)}$, replacing now

(III.32) by $\sqrt{n_2}\ \ \sqrt{n_\ell}$. (III.32')

(Since all pucks in $\Delta^{(1)}$ have already been chosen, $\ell \in \{2,\ldots,M\}$, so that $\sqrt{n_2}\,\sqrt{n_\ell}\ \geq n_\ell$); etc. All N resulting possibilities of forming dumbells can be labelled by planar diagrams $\tilde{\mathcal{G}}$; (each $\tilde{\mathcal{G}}$ is a family of P/2 line segments $\ell(\Delta,\Delta')$ joining two points in Δ, Δ', resp.). As our final estimate we now obtain, (see (III.31), (III.32), (III.32'), etc.):

$$N \leq \left[\prod_{\ell=1}^{M} n_\ell^{\frac{n_\ell}{2}}\right] \sup_{\tilde{\mathcal{G}}} \left[\prod_{\substack{\text{all dumbells} \\ \text{in } \tilde{\mathcal{G}}}} \text{length (dumbell)}^3\right] \tag{III.33}$$

If we have understood this well it is no surprise for us that

$$I \leq \left[\prod_{\ell=1}^{M} n_\ell^{\frac{n_\ell}{2}}\right] \sup_{\mathcal{G}} \left[|G| \prod_{\substack{\text{all internal} \\ \text{lines } \ell(\Delta,\Delta') \\ \text{of } \mathcal{G}}} d(\Delta,\Delta')^3\right],$$

or, in view of ((III.28) and) (III.29)

$$I \leq \left[\prod_{j \in \tilde{\Lambda}} (4p_{n_j})^{2p_{n_j}}\right]$$

$$\times \sup_{\mathcal{G}} \left[|G| \prod_{\substack{\text{all internal} \\ \text{lines } \ell(\Delta,\Delta') \\ \text{of } \mathcal{G}}} d(\Delta,\Delta')^3\right] \tag{III.34}$$

Given inequality (III.34), the proof of Lemma III.3 is now relatively easy: One uses (III.13) (<u>rapid fall off of $\langle\Phi_\kappa(x)\Phi_{\kappa'}(y)\rangle_0$</u>) to kill (absorb) the factors $d(\Delta,\Delta')^3$. As is easy to see, one is then lead to estimates ("on small graphs") of the type of Lemma III.1, (a). (The details are hardly more difficult than Exercise 9. See also [DG,EMS,MS]).

This completes our discussion of Theorem III.2 and Lemma III.3.

Remarks:

1) I am very much indebted to J. Magnen and R. Sénéor for patiently explaining to me the use of combinatoric factors in estimating Feynman diagrams and a proof of Lemma III.2 of which the above is a popularized account.
2) The proofs of [N2,GRS5] avoid the combinatorics explained above by means of more analysis. This more in analysis does however not extend to models in three space-time dimensions, whereas the type of methods explained here does, (mutatis mutandis).

Some remarks on $\lambda\phi_3^4$:

(a) For d=3

$$<\Phi_\kappa(x)^2>_0 = (2\pi)^{-3} \int \frac{\eta(k/\kappa)^2 d^3k}{k^2+m^3}$$

$$= O(\kappa)$$

(as opposed to $O(\log\kappa)$, for d=2).

(b) From (a) and (III.9) (definition of Wick ordering) one concludes that $:\Phi^4:(f)$ is not a well defined random variable;

$$||U_{\kappa,\Lambda}||_2^2 = O(\kappa).$$

(c) One is therefore forced to introduce counterterms to cancel ultraviolet divergences. (Since one wants to achieve the construction of a quantum measure that satisfies the hypotheses of Theorem II.3, in particular O-S positivity, these counterterms must be local). On the level of Feynman perturbation theory the only divergent diagrams are

cancelled by a (scalar) vacuum counterterm $E_{\kappa,\Lambda}$.

[diagram: line labelled κ with loop, labelled κ], which is cancelled by a mass counter-term $\delta m_\kappa^2 \int_\Lambda :\Phi_\kappa(x)^2: d^3x$, where

$$\delta m_\kappa^2 = O(\log\kappa).$$

(d) One may therefore define a renormalized Euclidean action

$$U^R_{\kappa,\Lambda}(\Phi) = U_{\kappa,\Lambda}(\Phi) + \delta m_\kappa^2 \int :\Phi_\kappa(x)^2: d^3x + E_{\kappa,\Lambda} \qquad \text{(III.35)}$$

Exercise 10: Show that each term in the perturbation series expansion in $U^R_{,}$ for e.g. the EGF's is well defined and finite. Combining (III.35) with (III.9), (a) and (c) we obtain

$$U_{\kappa,\Lambda} \geq -O(\kappa^2) \text{ (as opposed to } O(\log\kappa)^2), \text{ for } d=2),$$

$$\text{i.e. } e^{-U^R_{\kappa,\Lambda}} \leq e^{O(\kappa^2)} \qquad \text{(III.36)}$$

Fact: The methods used in the proof of Theorem III.2 break down in the case of $\lambda\phi^4_3$. The main reason is (III.36), (i.e.) the replacement of $e^{O((\log\kappa)^2)}$ by $e^{O(\kappa^2)}$). Glimm and Jaffe [GJ3] (see [MS] for subsequent simplifications) have found an ingenious refinement of the basic strategy used in the proof of Theorem III.2 to show that

$$\int e^{-U_{\kappa,\Lambda}(\Phi)} d\mu_0(\Phi) \leq e^{O(|\Lambda|)}, \qquad \text{(III.37)}$$

uniformly in κ.

Their methods involve expressing $U^R_{\kappa,\Lambda}$ as a sum of terms localized in phase space and doing then a truncated perturbation expansion of the l.h.s. of (III.37) to exhibit ultraviolet cancellations. Their analysis is very difficult.

III.3 The $\lambda\phi_2^4$ Theory: The Thermodynamic Limit

There are at least three different routes to construct a Euclidean invariant limit $d\mu_U$ (a quantum measure) of the family $\{d\mu_{U_\Lambda}\}$ (and of all the EGF's). Here we just mention these routes without going into any details.

(a) Gaussian) Expansions; [GJS1,2,3,4, Sp1,3]
They are the field theory analogue of high and low temperature expansions in statistical mechanics. The basic ingredients for these expansions are:

(i) The kernel of the covariance of $d\mu_0$ decreases exponentially fast;(it is the inverse of a differential operator).

(ii) U_Λ is local; i.e. the coupling of distant regions is entirely due to $d\mu_0$; the "effective" coupling constant is small; and, more technically:

(iii) Variants of Theorem III.2 and Lemma III.3 + a lot of difficult combinatorics

The expansion methods yield the most detailed information (e.g. detailed properties of spec (H,P); see [GJS 3,4 Sp 1,2]). Their drawback is that they have only a finite radius of convergence.

(b) Correlation inequalities; [GRS 3,4,5, N4]
These are the field theory analogue of Ising model methods of Griffiths. They do in general not yield detailed information about the limiting theory, but there is no restriction on the size of the coupling constant. These methods are based on the lattice approximation of $\lambda\phi^4$, [GRS 3], a technique which has proven to be very useful in many other contexts; see also [E,S1].

(c) FKG and Lee-Yang methods; [F5,FS].
These are based on combining the expansion methods of [Sp3] with correlation inequalities and the Lee-Yang theorem [S2,Nw]. They are inspired by Ising techniques of [LP], [LM] and others. They yield intermediate information (between (a) and (b); e.g. on the physical mass gap and absence of phase transitions) and are sometimes applicable in situations where (a) and (b) give no results. No restrictions on the size of the coupling constant are imposed. For an excellent account of (a) and (b) see the contributions of Glimm-Jaffe-Spencer, Guerra-Rosen-Simon and Nelson to [E]. All three methods also work for $\lambda\phi_3^4$; [ϕ_3^4]. They have the nice property that (once $\{d\mu_{U_\Lambda}\}$ has been constructed) they "auto-

matically" imply the hypotheses of Lemmas II.1 (=> O-S positivity) and II.2 (=> Euclidean invariance of $d\mu_U$). Moreover when combined with Theorem III.2, (b) (stability) they yield the exponential bounds of hypothesis (C) of the reconstruction theorem I.3. Therefore the limiting measures obtained in (a)-(c) are quantum measures from which a unique r.q.f.t. can be reconstructed.

IV. UNIQUENESS AND DEGENERACY OF THE PHYSICAL VACUUM AND THE CRITICAL POINT IN $\lambda\phi^4_{2,3}$

In this (final) section we review (without proofs) many of the rigorous results concerning the (uniqueness or) degeneracy of the physical vacuum (phase transitions) and the critical point in the $\lambda\phi^4$ theory in d=2 and 3 space-time dimensions. Our discussion is presented in terms of a "thermodynamic function", the infinite volume vacuum energy density of the $\lambda\phi^4$-theory (defined below; see [Gu,GRS5,S2,E,FS,GJ7], etc.) This is the field theory analogue of the free energy in statistical mechanics.

For the purposes of this section we must consider a (renormalized) Euclidean action U that includes quadratic and linear couplings. Formally

$$U_\Lambda = \int_\Lambda dx\{\lambda:\Phi(x)^4:-\frac{\sigma}{2}:\Phi(x)^2:-\mu\Phi(x)\} \tag{IV.1}$$

(+ counterterms, if d=3).

Here $\lambda>0$, σ and μ real, (and the counterterms are the ones used in Section III.2, (III.35)). Note that the counterterms can be chosen to be independent of σ and μ and, for d=3, Wick order :-: can be done with respect to bare mass 0; e.g.

$$:\Phi(0)\Phi(x): = \Phi(0)\Phi(x) - \frac{1}{4\pi|x|} \tag{IV.2}$$

These observations are important in the proof of existence of phase transitions in $\lambda\phi^4_3$ given in [FSS]. For d=2, Wick order with respect to bare mass 0 is impossible, and this is the basic, technical reason why phase transitions are more subtle in two dimensions, [GJS 4]!

Next we define - in the sense of Theorem III.2, (b) and its ϕ^4_3-version, [GJ3] -

$$N_\Lambda(\sigma,\mu) \equiv \int_{\mathscr{S}'} d\mu_o(\Phi)\; e^{-U_\Lambda(\Phi)} ; \qquad (IV.3)$$

(λ and the bare mass m in $d\mu_o$ being kept fixed throughout the rest of Section IV). We let <-> (σ,μ) denote expectation with respect to the $\phi^4_{2,3}$ quantum measure $d\mu_U$; (see Section III.3; and $[P(\Phi)_2, \phi^4_3]$ for complete proofs of existence of $d\mu_U$).

The first result concerns the existence of the thermodynamic limit of the vacuum energy density.

<u>Theorem IV.1:</u>
(a) $\alpha_\infty(\sigma,\mu) \equiv \lim_{\Lambda \nearrow \mathbb{R}^d} \frac{1}{|\Lambda|} \log N_\Lambda(\sigma,\mu)$ exists. ($\alpha_\infty(\sigma,\mu)$ is the "vacuum energy density" of the theory).
(b) $\alpha_\infty(\sigma,\mu)$ is the restriction of a function (also denoted α_∞) holomorphic in the complex μ-plane, except for a set of points on the imaginary axis, to the real μ-axis.

Proofs of (a) are given in [Gu] (ϕ^4_2) and [SeS,Pa] (ϕ^4_3); (b) is the ϕ^4-version of the <u>Lee-Yang theorem</u> and is due to [S2].

Next we give a necessary and sufficient condition for uniqueness (resp. degeneracy) of the physical vacuum in terms of α_∞.

<u>Theorem IV.2:</u>
(a) The physical vacuum Ω (see Section I) is <u>unique</u> if and only if $\alpha_\infty(\sigma,\mu)$ is continuously differentiable; in particular it is <u>unique</u>, for all $\mu \neq 0$.
(b) The one point function of the infinite volume $(\lambda\phi^4 - \frac{\sigma}{2}\phi^2 - \mu\phi)$-theory is given by

$$S_1 \equiv \langle\Phi(x)\rangle(\sigma,\mu) = \frac{\partial}{\partial\mu}\alpha_\infty(\sigma,\mu), \qquad (IV.4)$$

and the "<u>susceptibility</u>" χ by

$$\int \langle\Phi(o)\Phi(x)\rangle\ (\sigma,\mu)\ dx = \frac{\partial^2}{\partial\mu^2}\alpha_\infty\ (\sigma,\mu). \qquad (IV.5)$$

Part (a) is based on Theorem IV.1, (b) and is due to B. Simon (and R. Griffiths [S2]; see B. Simon's contribution to [E] and refs. given there). A more general and deeper result (inspired by [LM]) has been proven in [FS]. The first part of (b) is due to [S2] (see also [E]), the second part due to [F1], Section 3; (a more general result can be found in [F5], where the EGF's were shown to be derivatives of a generalized vacuum energy density and to have the same analyticity properties in μ as $\alpha_\infty(\sigma,\mu)$).

Definition:

$$\phi_c(\sigma) \equiv \langle\Phi(x)\rangle\ (\sigma,0+) = \frac{\partial}{\partial\mu}\alpha_\infty(\sigma,\mu)\Big|_{\mu=0+} \qquad \text{(IV.6)}$$

is called the "spontaneous magnetization." Using the $\Phi\to-\Phi$ symmetry of $d\mu_0$ and U (for $\mu=0$) one sees that

$$\langle\Phi(x)\rangle\ (\sigma,0-) = -\langle\Phi(x)\rangle\ (\sigma,0+) = -\phi_c(\sigma)$$

It has been proven by Guerra (see his contribution to [Bi]), for d=2, and in [GJ6], for d=3, that the r.q.f.t's reconstructed from

$$\langle-\rangle\ (\sigma,0\pm) \equiv \lim_{\pm\mu\searrow 0} \langle-\rangle\ (\sigma,\mu)$$

have a unique physical vacuum, (see also [FS]).
By Theorem IV.2, (a)

$$\langle-\rangle\ (\sigma,0+) \neq \langle-\rangle\ (\sigma,0-) \qquad \text{(IV.7)}$$

if and only if $\phi_c(\sigma) > 0$; (existence of a phase transition!)
For $\phi_c(\sigma)>0$, the $\Phi\to-\Phi$ symmetry of $d\mu_0$ and U is therefore spontaneously broken by the vacua reconstructed from $\langle-\rangle\ (\sigma,0\pm)$.
The physical mass (-gap) of the r.q.f.t. reconstructed from $\langle-\rangle\ (\sigma,0+)$ (= mass of the r.q.f.t. obtained from $\langle-\rangle\ (\sigma,0-)$!) is denoted $m_*(\sigma)$.

Theorem IV.3:
(a) $\phi_c(\sigma)$ is a non-decreasing function of σ.
(b) $m_*(\sigma)$ is a non-increasing function of σ, for all $\sigma\in\{\sigma':\phi_c(\sigma')=0\}$ (see Fig. 1,2,3).

Both (a) and (b) are straightforward consequences of a correlation inequality (the second Griffiths inequality; see [GRS3]).
Theorem IV.3 leads to the following

Definition:

$$\bar{\sigma}_c = \inf \ \{\sigma : \phi_c(\sigma) > 0\}$$

$$\underline{\sigma}_c = \sup \ \{\sigma : m_*(\sigma) > 0\};$$

The next result asserts existence of phase transitions.

Theorem IV.4: In d=2 and 3 space-time dimensions

(a) $-\infty < \underline{\sigma}_c \leq \bar{\sigma}_c < \infty$;

i.e. there is a phase transition, in the sense explained in (IV.7), as σ is increased beyond $\bar{\sigma}_c$.

(b) For all $\sigma \in (-\infty, \underline{\sigma}_c]$,

$$\frac{dm_*(\sigma)^2}{d\sigma} \leq 1,$$

and $m_*(\sigma) \searrow 0$, as $\sigma \nearrow \underline{\sigma}_c$, continuously.

That $\bar{\sigma}_c$ is finite (the main result!) is proven for d=2 in [GJS4] and for d=3 in [FSS]. (See also [F3] for a review).
That $\underline{\sigma}_c \leq \bar{\sigma}_c$ holds is proven in [MR], and that $\underline{\sigma}_c > -\infty$ is due [GJS 2,3]. The first part of (b) is contained in [GJ7] and refs. given there, the second part in [MR].

Theorem IV.5:

(a) for d=3, $m_*(\underline{\sigma}_c) = 0 = \phi_c(\underline{\sigma}_c)$, i.e. there exists a $\lambda\phi_3^4$-theory with a unique vacuum invariant under $\phi \to -\phi$, the physical mass (-gap) of which vanishes. For d=2 the same is true if one assumes that $\phi_c(\underline{\sigma}_c) = 0$.

(b) For $\sigma = \underline{\sigma}_c$, the integral

$$\chi(\underline{\sigma}_c) \equiv \int \langle \Phi(o)\Phi(x) \rangle \quad (\underline{\sigma}_c, 0) \ dx$$

diverges; ("divergent susceptibility").

(c) For d=2, $m_*(\sigma) > 0$ if $\sigma >> \bar{\sigma}_c$.

(d) If $\phi(\bar{\sigma}_c +) = 0$ (i.e. no "Thouless effect") then $m_*(\sigma) \searrow 0$, as $\sigma \searrow \bar{\sigma}_c$.

Part (a) has been proven in [GJ6] (where the results of $[\phi_3^4]$ and Theorem IV.4, (a) were assumed). See also the contribution of T. Spencer and the author to [Ca] for a review and more details. Reasonable, conjectured assumptions would imply that

$$\langle\Phi(o)\Phi(x)\rangle\ (\underline{\sigma}_c) \underset{|x|\to\infty}{\approx} O(|x|^{-(1+\eta)}),$$

with $0<\eta\leq 2$, (and $\eta\leq 2$ follows from part (b) of the Theorem) and furthermore that the scaling limit of the theory at $\sigma=\underline{\sigma}_c$ exists and defines a non-trivial r.q.f.t. with vanishing mass-gap, but <u>without</u> 0-mass particles!

Part (b) is due to [GJ 7]; (c) is a deep result of [GJS 4] (based on a difficult Gaussian expansion about mean field theory).

Part (d) has been shown in [F1] (using correlation inequalities). The "loop expansion" (see e.g. [Ca]) suggests that $\phi_c(\bar{\sigma}_c)=0$.

The following graphs summarize some knowledge about phase transitions in $\lambda\phi^4_{2,3}$; (the behaviour in the dashed regions is conjectured):

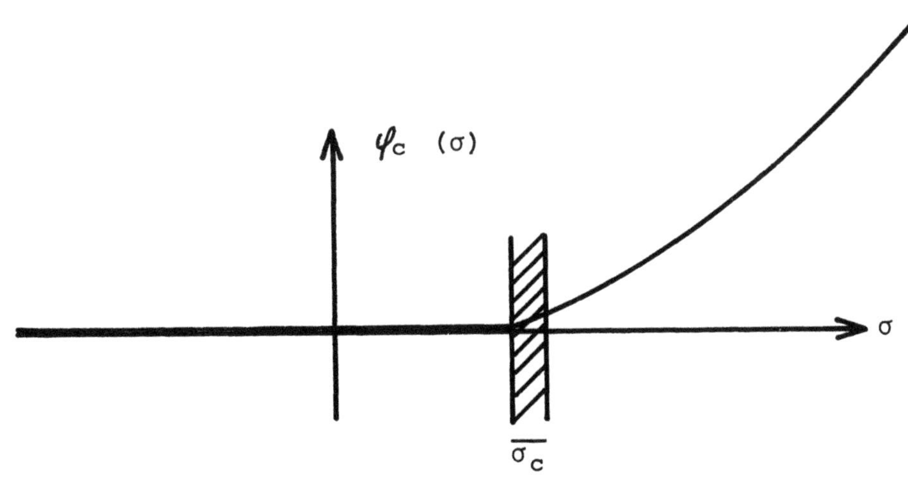

Figure 1. d = 2,3

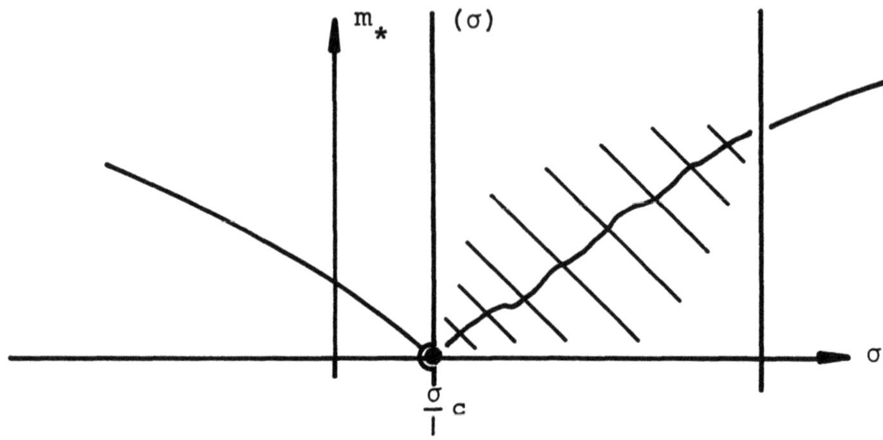

Figure 2. $d = 2$

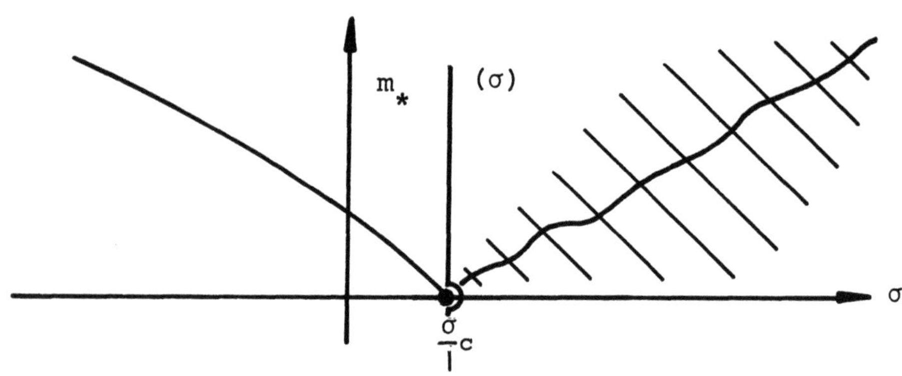

Figure 3. $d = 3$

<u>Other interesting results:</u>

1. Particles in $\phi^4_{2,3}$:

It is shown in [Sp 4] that, for $\sigma<\bar{\sigma}_c$, there are <u>no</u> two particle bound states (in the sense explained in [GJS1, Sp 4]). In [GJ7] it is concluded that, for all σ in the set $\{\sigma:\sigma\leq\underline{\sigma}_c, \frac{dm_*(\sigma)^2}{d\sigma}>0\}$, <u>there exists a discrete one</u>

particle shell in spec (H,P) of mass $m_*(\sigma)$.

2. Quantum Solitons in ϕ_2^4:

For d=2 the existence of two states <-> $(\sigma, 0+)$ and <-> $(\sigma, 0-)$ implies the existence of two disjoint, Poincaré-covariant superselection (soliton) sectors that are disjoint from ("orthogonal to") the two vacuum sectors. This is proven in [F4]. (See [F3] for a review, and [Co,R,St] for other results and references). Preliminary results indicate that, for $\sigma >> \sigma_c$, the mass gap m_s on the soliton sectors is positive*) and that there are particles (quantum (anti-) solitons) of mass m_s; (see [F3]; and [Co] for more heuristic results).

3. Phase transitions and Goldstone bosons, d=3:

For the $[\lambda(\vec{\phi}\cdot\vec{\phi})^2 - \frac{\sigma}{2}(\vec{\phi}\cdot\vec{\phi})]_3$-theory, with $\vec{\phi}=(\phi_1,\ldots,\phi_N)$, N>1, (i.e. multi-component ϕ_3^4) existence of phase transitions and spontaneous O(N) symmetry breaking has been proven in [FSS]. For N=2,3 it follows (up to a minor technical estimate) that there exist N-1 Goldstone bosons, (0-mass one particle states).

Concluding remarks: With the results on $\lambda\phi_3^4$ and $\lambda(\vec{\phi}\cdot\vec{\phi})_3^2$ mentioned throughout the text one has come close- if not to realistic r.q.f.t.'s - to what real physicists may be interested in. This is because these models, at their critical point, are supposed to describe the scaling limit (long distance behaviour) of physical systems (such as ferromagnetic crystals) at the critical temperature; (see e.g. [Ca] for a review). Yet, many desirable results are still missing.

I have not touched in these notes the problem of "Fermions in c.q.f.t.", i.e. the construction and analysis of models including Fermions in their particle spectrum. For this, see [GJ1] (Hamiltonian formalism), [Y_2] (Euclidean description), where the Yukawa interaction has been analyzed in detail. (see also [Co,FSe,Lü,St] , and refs. given there for special results and two dimensional peculiarities). I hope that at the next conference in Bielefeld (that concerns quantum physics) somebody will speak about existence and non-triviality (or is it trivial?) of $\lambda\phi_4^4$. Some c.q.f.t. insight into this theory has already been gained; see the contributions of Glimm and Jaffe to [Bi,Ca,M] (and refs. given there) and Schrader's proposals, [Sc]. But it is still very insufficient. My secret suspicion is however that, at a future conference, it will be gauge theories that dominate

*) This has recently been proven!

c.q.f.t.. Finally I wish to express my modest hope that these notes may contribute to a little more appreciation of c.q.f.t. among people who regard themselves as non-experts.

REFERENCES

[Bi] Proceedings of the conference on Quantum Dynamics: Models and Mathematics, Bielefeld 1975; to appear.

[BW] J.Bisognano and E. Wichmann, J. Math. Phys. 16, 985, (1975).

[BY] H.J.Borchers and J.Yngvason, "Necessary and Sufficient Conditions for Integral Representations of Wightman Functionals at Schwinger Points", preprint 1975; see also Borchers' contribution to [M].

[BEG] J.Bros, H.Epstein and V.Glaser, Helv.Phys. Acta 45, 149, (1972).

[B] C.Burnap, "Isolated One Particle States in Boson Quantum Field Theory Models", Harvard University preprint 1975. (see also J.Glimm and A.Jaffe, Phys.Rev. DII, 2816, (1975) and ref. [Sp 1]).

[Ca] Proceedings of the summer school in theoretical physics, Cargèse, Corsica, 1976; to appear.

[Co] S.Coleman, "Classical Lumps and Their Quantum Descendants", to appear in the proceedings of the 1975 International School on Subnuclear Physics "Ettore Majorana", (Erice).

[D1*] J.Dimock, Commun.math.Phys. 35, 347, (1974); see also his contribution to [E].

[D2] J.Dimock, "The $P(\Phi)_2$ Green's Functions: Asymptotic Perturbation Expansion", Buffalo preprint 1975, to appear.

[DE*] J.Dimock and J.-P.Eckmann,"Spectral Properties and Bound State Scattering for Weakly Coupled $\lambda P(\Phi)_2$ Models", ZiF-University of Bielefeld, preprint 1976, to be published.

[DG*] J.Dimock and J.Glimm, Adv.Math. 12,58, (1974).

[DHR] S.Doplicher, R.Haag and J. Roberts, Commun.math. Phys. 23, 199, (1971) and 35, 49, (1974).

[DF] W.Driessler and J.Fröhlich, "The Reconstruction of Local Observable Algebras from the Euclidean Green's Functions of a Relativistic Quantum Field Theory", ZiF, Univ. Bielefeld, Preprint 1976, to be published.

[EEF*] J.-P.Eckmann, H.Epstein and J.Fröhlich, "Asymptotic Perturbation Expansion for the S-Matrix and ...", Ann. Inst. H. Poincaré, 25, 1, (1976).

[EMS*] J.-P. Eckmann, J. Magnen and R. Sénéor, Commun. math. Phys. 39, 251, (1975).

[E*] G. Velo and A. Wightman (eds.), "Constructive Quantum Field Theory", Lecture Notes in Physics 25, (1973), Springer, Berlin-Heidelberg-New York. (Proceeding of "Erice, 1973").

[F1] J. Fröhlich, Ann.Phys. 97, 1, (1976).

[F2*] J. Fröhlich, Helv.Phys.Acta 47, 265, (1974), and "Schwinger Functions and Their Generating Functionals, II..." to appear in Adv. Math.

[F3] J. Fröhlich, "Phase Transitions, Goldstone Bosons and Topological Superselection Rules" in "Current Problems in Elementary Particle and Mathematical Physics", P. Urban (ed.), Springer, Wien-New York, 1976.

[F4*] J. Fröhlich, Commun. math.Phys. 47, 269, (1976).

[F5*] J. Fröhlich, contribution to ref. [M], and "Existence and Analyticity in the Bare Parameters of the $\lambda(\vec{\phi}\cdot\vec{\phi})^2$-Quantum Field Model", (report, 1975; so far unpublished).

[FSe] J. Fröhlich and E. Seiler, "The Massive Thirring Schwinger Model (QED_2): Convergence of Perturbation Theory and Particle Structure", to appear in Helv. Phys. Acta, 1976.

[FS*] J. Fröhlich and B. Simon, "Pure States for General $P(\Phi)_2$ Theories: Construction, Regularity and Variational Equaliy", to appear in Ann. Math.

[FSS] J. Fröhlich, B. Simon and T. Spencer, "Infrared Bounds, Phase Transitions and Continuous Symmetry Breaking", Commun.math.Phys. 50, 79, (1976), (see also Phys. Rev. Letters 36, 804, (1976)).

[Gl] V. Glaser, Commun.math.Phys. 37, 257, (1974).

[G] J. Glimm, Commun.math.Phys. 8, 12, (1968).

[GJ1] J. Glimm and A. Jaffe, in "Statistical Mechanics and Quantum Field Theory", Les Houches 1970, C.de Witt and R. Stora (eds.), Gordon and Breach, New York, 1971.

[GJ2] J. Glimm and A. Jaffe, in "Mathematics of Contemporary Physics", R. Streater (ed.), Academic Press, New York, 1972.

[GJ3] J. Glimm and A. Jaffe, Fortschritte d. Physik 21, 327, (1973).

[GJ4] J. Glimm and A. Jaffe, Acta Math. 125, 203, (1970).

[GJ5] J. Glimm and A. Jaffe, J. Math. Phys. 13, 1568, (1972).

[GJ6] J. Glimm and A. Jaffe, Ann.Inst. H. Poincaré, 22, 109, (1975); see also Ann. Inst. H. Poincaré 22, 97, (1975).

[GJ7] J. Glimm and A. Jaffe, "Critical Exponents and Elementary Particles", Rockefeller University, preprint 1976, to be published.

[GJ8] J. Glimm and A. Jaffe, see [GJ4] and Commun. Math.Phys. 22, 253, (1971); also [GJ2].

[GJS1] J. Glimm, A. Jaffe and T. Spencer, see their contribution to [E], Part I.

[GJS2*] J. Glimm, A.Jaffe and T. Spencer, see their contribution to [E], Part II; ("Cluster Expansion").

[GJS3*] J. Glimm, A. Jaffe and T. Spencer, Ann.Math. 100, 585, (1974).

[GJS4*] J. Glimm, A. Jaffe and T. Spencer, Commun. math.Phys. 45, 203, (1975) and "A Convergent Expansion about Mean Field Theory", (Part I and II), to appear in Ann.Phys.

[Gu*] F. Guerra, Phys. Rev. Letters 28, 1213, (1972).

[GRS1*] F. Guerra, L. Rosen and B. Simon, Commun.math. Phys. 27, 10, (1972).

[GRS2*] F. Guerra, L. Rosen and B. Simon, Commun.math. Phys. 29, 233, (1973).

[GRS3*] F. Guerra, L. Rosen and B. Simon, Ann.Math. 101,111, (1975).

[GRS4*] F. Guerra, L. Rosen and B. Simon, Commun.math. Phys. 41, 19, (1975).

[GRS5] F. Guerra, L. Rosen and B. Simon, "Boundary Conditions in the $P(\Phi)_2$ Euclidean Field Theory", to appear in Ann.Inst. H. Poincaré, 1976.

[He] K. Hepp, "Théorie de la Renormalisation", Lecture Notes in Physics 2, Springer Verlag, Berlin-Heidelberg-New York, 1969.

[HK] R. Haag and D. Kastler, J. Math. Phys. 5, 848, (1964).

[J] R. Jost, "The General Theory of Quantized Fields", Ann.Math. Soc. Publ., Providence, R.I., 1965.

[LM] J. Lebowitz and A. Martin-Löf, Commun.math. Phys. 25, 276, (1972).

[LP] J. Lebowitz and O. Penrose, Commun.math. Phys. 11, 99, (1968).

[Lü] M. Lüscher, "Dynamical Charges in the Quantized, Renormalized Massive Thirring Model", DESY preprint, (Hamburg), 1976.

[MS] J. Magnen and R. Sénéor, "L'Interaction ϕ^4 à trois Dimensions: Positivité de l'Hamiltonien, Développement à Faible Constante de Couplage et Analyticité" Orsay 1976, C.N.R.S. document A.O. 12.829.

[M] "Les Méthodes Mathématiques de la Théorie Quantique des Champs", Editions du C.N.R.S., Paris, 1976; (proceedings of the 1975 Colloquium in Marseille).

[N1] E. Nelson, in "Mathematical Theory of Elementary Particles", R. Goodman and I. Segal, (eds.), M.I.T. Press, Cambridge, Mass. 1966.

[N2*] E. Nelson, in "Partial Differential Equations", D. Spencer (ed.), Symp. in Pure Math. Vol. 23, A.M.S. Publ. 1973.

[N3] E. Nelson, J. Funct. Analysis 12, 97, (1973).

[N4*] E. Nelson, his contribution to [E].

[N5] E. Nelson, J. Funct. Analysis 11, 211, (1972).

[Nw] C. Newman, his contribution to [E] and [M] and refs. given there.

[O] K. Osterwalder, his contribution to [E].

[OS] K. Osterwalder and R. Schrader, Commun. math. Phys., 31, 83, (1973), and 42, 281, (1975).

[Pa] Y. Park, "Uniform Bounds of the Schwinger Functions in Boson Field Models", to appear in J. Math. Phys.

[$P(\Phi)_2$] refs marked with *

[R] J. Roberts, "Local Cohomology and Superselection Structure", C.N.R.S. Marseille preprint 1976.

[Ru] D. Ruelle, Nuovo Cimento 19, 356, (1961).

[S≠1] see [EEF], [D2] and K. Osterwalder and R. Sénéor "The Scattering Matrix is Non-Trivial for Weakly Coupled $P(\Phi)_2$ Models", to appear in Helv. Phys. Acta, 1976, also Eckmann's and Osterwalder's contribution to [Bi].

[Sc] R. Schrader, "A Possible Constructive Approach to ϕ_4^4, I-III", preprints 1975-76, to be published; his contribution to [M].

[SeS] E. Seiler and B. Simon, "Nelson's Symmetry and All That in the (Yukawa)$_2$ and $(\phi^4)_3$ Field Theories", to appear in Ann. Phys. 1976.

[S1*] B. Simon, "The $P(\Phi)_2$ Euclidean (Quantum) Field Theory", Princeton University Press, Princeton N.J., 1974.

[S2*] B. Simon and R. Griffiths, Commun.math.Phys. 33, 145, (1973).

[Sp1*] T. Spencer, Commun.math.Phys. 44, 143, (1975).

[Sp2*] T. Spencer and F.Zirilli, Commun.math.Phys. 49, (1976).

[Sp3*] T. Spencer, Commun.math.Phys. 39, 63, (1974).

[Sp4] T. Spencer, Commun.math.Phys. 39, 77, (1974).

[St] R. Streater and I. Wilde, Nucl. Phys. B24, 561, (1970), R. Streater, Acta Physica Austriaca, Suppl. XI, 317, (1973), P. Bonnard and R. Streater, Helv. Phys. Acta 49, (1976).

[SW] R. Streater and A.S. Wihgtman, "PCT, Spin and Statistics and All That", Benjamin, New York, 1964.

[Sy1] K. Symanzik, "A Modified Model of Euclidean Quantum Field Theory", NYU Report, 1964.

[Sy2] K. Symanzik, J.Math.Phys. 7, 510, (1966).

[Sy3] K. Symanzik, in "Local Quantum Theory", R. Jost (ed.), Academic Press, New York, 1969.

[W1] A. Wightman, in "Fundamental Interactions in Physics and Astrophysics", B.Kursunoglu et al. (eds.), Plenum Press 1973.

[Y_2] Hamiltonian formalism: see [GJ1] and refs. given there.
Euclidean formalism:
E. Seiler, Commun.math.Phys. 42, 163, (1975).
E. Seiler and B. Simon, J.Math.Phys. (1976).
E. Seiler and B. Simon, Commun.math.Phys. 45 99, (1975).
E. Seiler and B. Simon, ref. [SeS].
O. McBryan, Commun.math.Phys. 44, 237, (1975).
O. McBryan, Commun.math.Phys. 45, 279, (1975).
O. McBryan, his contribution to [M].
J. Magnen and R. Sénéor, "The Wightman Axioms for the Weakly Coupled Yukawa Model in Two Dimensions", to appear in Commun.math.Phys., 1976.
A. Cooper and L. Rosen, "The Weakly Coupled $(\text{Yukawa})_2$ Field Theory; Cluster Expansion and Wightman Axioms", Princeton University, preprint 1976.
A semi-Euclidean formalism has been proposed and applied to Yukawa models by P. Federbush and D. Brydges, (J.Math.Phys., (1975)).

[ϕ^4_3] [GJ3] [MS] [F5] [Pa] [SeS],and
J. Feldman, Commun.math.Phys. 37, 93, (1974).
J. Feldman and K. Osterwalder, Ann.Phys. 97, 80, (1976).
J. Feldman and K. Osterwalder, their contribution to [M].
J. Magnen and R. Sénéor, Ann.Inst.H. Poincaré 24, 95, (1976)
Y. Park, J.Math.Phys. 16, 1065, (1975); in "Current Problems in Elementary Particle and Mathematical Physics", P. Urban, (ed.), Springer-Verlag, Wien-New York, 1976.

ASYMPTOTIC BEHAVIOR OF THE AUTOCOVARIANCE FUNCTION AND VIOLATION OF STRONG MIXING

M. Cassandro and G. Jona-Lasinio*

Istituto di Fisica and Gruppo GNSM,
Università Roma
Rome, Italy

ABSTRACT

We discuss violation of the strong mixing condition for a class of multidimensional stochastic processes which includes cases of physical interest like ferromagnetic systems. We give a sufficient condition for strong mixing violation in terms of the asymptotic behavior of the autocorrelation function over large distances. As an important example to which our criterion applies, we discuss the two dimensional Ising model at the critical point. Connections with other problems in statistical physics and probability theory are briefly reviewed.

1. INTRODUCTION

In the applications of probability theory to physics and other branches of natural sciences, the central limit theorem is generally expected to hold. This belief is usually based on considerations of the following type. Given a physical system consisting of a large number of interacting subsystems, any finite portion of it, due to the finite range of the forces, is essentially independent from the rest. This point of view is at the root

---

* One of the authors (G. J.-L.) would like to thank Prof. L. Streit for the kind hospitality received at ZiF.

of the probabilistic approach to statistical mechanics developed by Khinchin a few decades ago [1] in which the central limit theorem provides the key for the understanding of all the basic formulas which had been obtained by physicists on the basis of heuristic arguments. Actually the central limit theorem is valid under rather general conditions and it can be applied to dependent random variables provided this dependence is not too strong.[2]

Recent developments in statistical mechanics have shown however that the above picture cannot be always true. If a system undergoes a phase transition of second order, the hypothesis of weak dependence of dynamical variables cannot be applied at the critical point. This situation is revealed by the recent progress related to the use of the so called renormalization group techniques. Even if this approach is still in many cases mathematically non rigorous it definitely indicates that limit distributions describing simple physical systems at the critical point can be non gaussian.[3]

All this immediately raises the question of characterizing the degree of dependence of a set of random variables which produces such deviations from the central limit theorem and of determining the kind of limit distributions that can appear in situations of physical interest.

In the theory of stochastic processes it is customary to introduce a hierarchy of conditions which express different degrees of dependence of random variables. We may mention complete regularity, uniformly strong mixing, strong mixing, etc. [4] In the studies of limit theorems for dependent variables that one can find in the literature, the strong mixing condition plays an especially important role. It is in fact sufficient to guarantee that a large sum of random variables behaves as if these were independent. A very interesting theorem asserts that the limit distribution, if it exists, in such a case is either a gaussian or a stable distribution.[2] If we consider that stable distributions different from the gaussian have an infinite second moment, a circumstance which makes them not very reasonable for many physical applications, we immediately understand why the central limit theorem retains a major importance also in situations where the variables cannot be taken as independent also in first approximation. This is the case that one encounters in Ising systems far from the critical point for which it is known that correlations decay exponentially and that the central limit theorem is satisfied.[5]

As we mentioned earlier, the critical point seems to fall outside the range of validity of the previous theorem and the question naturally arises of proving rigorously that, at least for a certain class of critical systems, the strong mixing condition does not hold. The interest of such a result lies in the fact that it may provide, as suggested elsewhere [6], a characterization of criticality directly in terms of probabilistic concepts and shows in addition that important physical applications require a systematic extension of limit theorem to situations of "strong" dependence which are beyond those usually treated by the traditional approach.

In the present paper we shall prove that an important class of physical systems, namely ferromagnetic system, are characterized by violation of strong mixing if their autocorrelation function decays like a non integrable power law. Furthermore, by combining this theorem with known results about the 2-dimensional critical Ising model, we reach the conclusion that our description applies in particular to this system.

The problem of describing the limit distributions which can appear under violation of strong mixing is much more difficult. For the systems we consider, if these limit distributions exist, they are characterized by the fact that their characteristic functions are limits of sequences of entire functions with zeroes on the imaginary axis. As we will see, this implies that if they are not gaussians they cannot be infinitely divisible either.

2. THE STRONG MIXING CONDITION

We consider a random field X_t with $t \in Z^\nu$ defined by a probability measure on the σ-algebra generated by sets of the space $K=(R)^{Z^\nu}$ (R is the real line) of the form

$$\{X_{t_1} \in A_1, \ldots X_{t_n} \in A_n\} \quad t_1, \ldots t_n \in \Lambda$$

Λ is an arbitrary finite set in Z^ν, A_i are Borel sets on the real line. For Λ fixed we call Σ_Λ the corresponding σ-algebra.

Consider now two finite sets Λ_1 and $\Lambda_2 \in Z^\nu$. The distance between Λ_1 and Λ_2 can be defined by

$$d(\Lambda_1,\Lambda_2) = \min_{t_1\in\Lambda_1, t_2\in\Lambda_2} |t_1-t_2| \tag{1}$$

where $|t_1-t_2|$ is for example the Euclidean distance. Following Dobrushin (7) we introduce the notion of strong mixing in the following way. Set

$$\delta(\Lambda_1,\Lambda_2) = \sup_{A\in\Sigma_{\Lambda_1}, B\in\Sigma_{\Lambda_2}} |P(A\cap B)-P(A)P(B)| \tag{2}$$

A random field with distribution P is said to possess the property of strong mixing if

$$\delta(\Lambda_1,\Lambda_2) \leq \alpha(d(\Lambda_1,\Lambda_2)) \tag{3}$$

where $\alpha(d) \to 0$ as $d\to\infty$. α is called the mixing coefficient. For later purposes we need the following lemma which is an obvious generalization of theorem 17.2.1 of ref. (2).

<u>Lemma 2.1</u>: Necessary and sufficient condition for a random field to be strong mixing is that the following condition holds

$$\sup |E(\xi\eta)-E(\xi)E(\eta)| \leq 4\, C_1C_2\mu(d(\Lambda_1,\Lambda_2))$$

ξ measurable in Σ_{Λ_1} (4)
η measurable in Σ_{Λ_2}

$$|\xi| \leq C_1$$

$$|\eta| \leq C_2$$

where $\mu(d) \to 0$ as $d \to \infty$. E is the expectation.

<u>Proof</u> - Sufficiency
We take for ξ and η indicator functions of measurable sets $A \in \Sigma_{\Lambda_1}$ and $B \in \Sigma_{\Lambda_2}$. Then (4) reduces to (3).

Necessity - Using the properties of conditional expectations,[8] we have

$$|E(\xi\eta)-E(\xi)E(\eta)| = |E\{\xi[E(\eta|\Sigma_{\Lambda_1})-E(\eta)]\}| \leq$$

$$\leq C_1\ E|E(\eta|\Sigma_{\Lambda_1})-E(\eta)| = C_1\ E\{\xi_1[E(\eta|\Sigma_{\Lambda_1})-E(\eta)]\}$$

where $\xi_1 = \text{sign}\ \{E(\eta|\Sigma_{\Lambda_1})-E(\eta)\}$.

Clearly ξ_1 is measurable with respect to Σ_{Λ_1} and therefore

$$|E(\xi\eta)-E(\xi)E(\eta)| \leq C_1|E(\xi_1\eta)-E(\xi_1)E(\eta)|$$

Similarly we may compare η with

$$\eta_1 = \text{sign}\ \{E(\xi_1|\Sigma_{\Lambda_2})-E(\xi_1)\}$$

to give

$$|E(\xi\eta)-E(\xi)E(\eta)| \leq C_1C_2|E(\xi_1\eta_1)-E(\xi_1)E(\eta_1)|$$

Introducing the events

$$A = \{\xi_1 = 1\} \in \Sigma_{\Lambda_1}$$
$$B = \{\eta_1 = 1\} \in \Sigma_{\Lambda_2}$$

the strong mixing condition (3) gives

$$|E(\xi_1\eta_1) - E(\xi_1)E(\eta_1)| =$$

$$|P(A\cap B)+P(\bar{A}\cap\bar{B})-P(\bar{A}\cap B)-P(A\cap\bar{B})$$

$$-P(A)P(B) - P(\bar{A})P(\bar{B})+P(\bar{A})P(B)+P(A)P(\bar{B})| \leq$$

$$\leq 4\ \alpha(d(\Lambda_1,\Lambda_2))$$

q.e.d.

3. FERROMAGNETIC SYSTEMS WITH PAIR INTERACTIONS

Let us now give some definitions used in the remainder of this article. A finite ferromagnetic Ising model is a triple (Λ,H,ρ) where

(1) Λ is a set of sites as in the previous section and we associate with each site $t\in\Lambda$ a spin variable X_t. The space $K=(R)^{Z^\nu}$ is called the phase space.

(2) The Hamiltonian H in the case of pair interactions, has the following form

$$H(X) = - \sum_{t,t' \in \Lambda} J_{t,t'} X_t X_{t'} \qquad J_{t,t'} \geq 0$$

Usually in the Hamiltonian one includes a magnetic field term of the form $\sum_{t\in\Lambda} h_t X_t$. In the following we shall be interested in the case h=0.

(3) The single spin measure ρ is an even Borel probability measure on R which decays sufficiently rapidly

$$\int_R e^{Q|x|^2} \rho(dx) \quad \forall Q \in R$$

and which may be approximated by spin $\frac{1}{2}$ distributions.

The Gibbs measure P of (Λ,H,ρ) is the measure on the phase space $(R)^\Lambda$ defined by

$$P_\Lambda(A) = \frac{\int_A e^{-\beta H(x)} \prod_{t\in\Lambda} \rho(dX_t)}{\int_{(R)^\Lambda} e^{-\beta H(x)} \prod_{t\in\Lambda} \rho(dX_t)} \qquad (1)$$

A is a measurable set in $(R)^\Lambda$. In statistical mechanics one is interested in the limit $\Lambda \to Z^\nu$ of the above expression. Expectations in the $\Lambda\to\infty$ limit will be denoted E(·).

For ferromagnetic systems with pair interactions the following interesting result has been established by Newman [(9)].

Theorem 3.1 $f(z) = E\left[e^{z(\sum_{t\in M\subset Z^\nu}(X_t - E(x_t)))}\right]$ where M is a finite set, is an entire function of z bounded by

$$e^{b_M|z|^2} \quad \text{where } b_M = \frac{1}{2} E\left[\left(\sum_{t\in M}(X_t - E(X_t))\right)^2\right]$$

Proof: By expanding f(z) around z=o we obtain

$$f(Z) = \sum_{o}^{\infty} \frac{z^{2n}}{(2n)!} E\left[\left(\sum_{t\in M}(X_t - E(x_t))\right)^{2n}\right]$$

Using gaussian inequalities (9)

$$E[(\sum_{t\in M}(X_t-E(X_t)))^{2n}] \leq \frac{(2n)!}{2^n n!}(E[(\sum_{t\in M}(X_t-E(X_t)))^2])^n$$

(2)
(3)

we find that the series expansion is absolutely convergent in every bounded subset of the complex plane and is therefore an entire function. Furthermore it is everywhere bounded by $e^{b_M|z|^2}$

q.e.d.

4. VIOLATION OF STRONG MIXING FOR FERROMAGNETIC SYSTEMS

For simplicity we first discuss the one dimensional case.

Theorem 4.1

A ferromagnetic system with pair interactions for which the autocorrelation function

$$E(X_o X_J)-E(X_o)E(X_J) = R(J) \quad \text{is such that}$$

$$\lim_{\tau\to\infty} \frac{\sum_{i=\tau}^{n\tau} R(i)}{\sum_{i=o}^{\tau} R(i)} \neq 0 \tag{1}$$

for n arbitrary, does not satisfy the strong mixing condition.

Proof: We will show that it is possible to construct local observables ξ and η such that the necessary condition for strong mixing given by Lemma 2.1 is not satisfied. Call

$$A_\tau^1 = \sum_{i=-\tau}^{o} \frac{X_i-E(X_i)}{B_\tau}$$

$$A_\tau^2 = \sum_{i=\tau}^{2\tau} \frac{X_i-E(X_i)}{B_\tau}$$

where $B_\tau^2 = E([\sum_{i=o}^{\tau}(X_i-E(X_i))]^2)$

and set

$$\xi = e^{i r A_\tau^1}$$

$$\eta = e^{i r A_\tau^2} \qquad r \text{ real}$$

obviously $|\xi| = |\eta| = 1.$

Consider now

$$E(e^{z(A_\tau^1 + A_\tau^2)})$$

$$E(e^{z A_\tau^1})$$

$$E(e^{z A_\tau^2})$$

By theorem 3.1 these are entire functions of z bounded for any τ by $e^{\frac{|z|^2}{2}}$. In particular they can be expanded around the origin. For r sufficiently small therefore we have

$$|E(\xi\eta) - E(\xi)E(\eta)| \geq \frac{r^2}{4} E(A_\tau^1 A_\tau^2) =$$

$$= \frac{r^2}{4} \sum_{i=-\tau}^{0} \sum_{j=\tau}^{2\tau} \frac{R(j-i)}{B_\tau^2} \to \frac{r^2}{4} \frac{\sum_{i=\tau}^{3\tau} R(i)}{\sum_{i=0}^{\tau} R(i)}$$

Under the assumptions of the theorem violation of strong mixing follows. The multidimensional case requires only some modification in the notations.

Denote with Λ_S^L the cube

$$\Lambda_S^L : L\, S^i \leq t^i < L(S^i+1) \quad i = 1,2 \ldots \nu$$

The analogue of (1) is then

$$\lim_{L\to\infty} \frac{\sum_{t_1 \in \Lambda_0^L} \sum_{t_2 \in \Lambda_S^L} R(t_2 - t_1)}{\sum_{t_1, t_2 \in \Lambda_0^L} R(t_2 - t_1)} =$$

$$\lim_{L\to\infty} \frac{\sum\limits_{L(s^i-1)\leq t^i<L(s^i+1)} R(t)}{\sum\limits_{o\leq t^i<L} R(t)} \neq o \qquad (2)$$

It is clear however that any sequence of pairs of sets for which (2) is satisfied when their distance goes to infinity will work equally well. This possibility will be exploited in the next section. We conclude this section with some remarks of a more general nature. It is obvious from our proof that for the conclusion to hold it is sufficient the existence of a third derivative with respect to z uniformly bounded in τ in some neighborhood of z=o of expectations of the form $E(e^{z\sum_1^\tau \frac{X_i}{B_\tau}})$. The interesting problem is to determine the class of physical systems which, besides ferromagnets with pair interactions, satisfy such a condition.

As a second comment, not unrelated to the previous one, we would like to emphasize that in general a condition like (4.1) or (4.2) is not sufficient to insure violation of strong mixing. A counter example, as discussed in [6] can be obtained from a model proposed by Davydov [10].

In principle we could also consider systems for which violation of strong mixing is due to higher order moments but of course these could not be ferromagnetic according to the definition adopted in this paper.

5. THE TWO-DIMENSIONAL ISING MODEL

As explained in the introduction our results are expected to become relevant for physics when applied to systems that are known to undergo a second order phase transition. It is standard in this field to examine first what happens in the case of the two-dimensional Ising model. It has been proven recently by Hegerfeld and Nappi [11] that the non critical Ising model is strong mixing. Actually there existed already evidence, based on limit theorems [5], that this property might hold. As far as the critical point is concerned, we will now show that using known results on the behaviour of critical correlation functions in conjunction with our theorem, it follows that the two-dimensional Ising model fails to be strong mixing at T_c. According to Wu and Mc Coy (12) the critical two spin correlation function $<\sigma_{oo}\sigma_{ii}>$ can be calculated in closed form and one finds the asymtotic behaviour $\frac{c}{i^{1/4}}$. By considering in this case the sums $\sum_{i=\tau}^{2\tau} <\sigma_{oo}\sigma_{ii}>$ and $\sum_{i=-\tau}^{o} <\sigma_{oo}\sigma_{ii}>$ our theorem implies violation of strong mixing. Therefore in this case the conjecture that violation of strong mixing characterizes the critical point holds true.

6. REMARKS ON LIMIT DISTRIBUTIONS

Having established that the spins of a critical 2-dimensional Ising model represent a collection of strongly dependent variables, the next important step is to calculate the limit distribution of normalized sums. To that purpose it would be enough to compute the $\lim_{\tau\to\infty} E(e^{iz \sum_i^\tau \frac{X_i}{B_\tau}})$ since the limit distribution is completely determined by its characteristic function. Unfortunately even in the case of ferromagnetic pair interactions it is not known whether such a limit exists in the critical case. If it exists however, we can immediately conclude from general theorems that if it is not a Gaussian it cannot be infinitely divisible. This can be seen as follows. Following Newman (13) we say that a random variable X is of type $\mathcal{L}$, if for some C and C^1 $|E(e^{ZX})| \leq C\, e^{C^1|Z|^2}$ for all Z and $E(e^{ZX})$ is even with only pure imaginary zeroes. For these variables Newman proved that

<u>Theorem 6.1</u>: If X is of type $\mathcal{L}$ then $E(e^{ZX}) = e^{bZ^2} \prod_j (1+\frac{Z^2}{\alpha_j^2})$ for some $b\geq 0$ and $0<\alpha_1\leq\alpha_2\leq\ldots$with $\sum_i \frac{1}{\alpha_j^2} <\infty$; the set $\{\alpha_j\}$ may be empty, finite or infinite.

There is a connection between ferromagnetic systems and variables of type $\mathcal{L}$ given by the following proposition.

Theorem 6.2: If the zeroes of $\int e^{ZX}\rho(dX)$ are pure imaginary, then for any choice of $\lambda_i \geq 0$ the zeroes of $E(e^{Z\sum_{i\in M}\lambda_i X_i})$, where M is a finite set, are pure imaginary.

The variables $\sum_i^\tau \frac{X_i}{B_\tau}$ for the Ising models are theorefore variables of type $\mathcal{L}$.

It is also easily seen that the zeroes $\alpha_i(\tau)$ cannot accumulate at the origin as $\tau\to\infty$. This follows from the identity valid for any τ

$$2(b(\tau)+\sum_i \frac{1}{\alpha_j(\tau)^2}) = 1$$

which implies $\alpha_i^2 > 1$, $\forall i$ -

From a general property of characteristic functions on the other hand we have that an expectation $E(e^{ZX})$ analytic at the origin is analytic in a strip containing the imaginary axis [(14)]. The characteristic function of the variables $\sum_i^\tau \frac{X_i}{B_\tau}$ in the limit $\tau\to\infty$ if it exists, is therefore an entire function which admits the representation of Theorem 6.1. The gaussian is then the only possible way to satisfy infinite divisibility[(*)]

(*) G. Hegerfeldt and C.Nappi pointed out to us the following argument which leads to conclusions similar to ours. A necessary condition for infinite divisibility is the positive definiteness of any truncated 2n-point function (G.Hegerfeldt, Comm.Math.Phys. 95 137 (1975)) - Since the Lebowitz inequality implies that the truncated 4-point function is non positive if h=0 (J.Lebowitz Comm.Math.Phys. 35, 87 (1974)) the only allowed infinitely divisible law is the Gaussian.

REFERENCES

(1) A.I.Khinchin, "Mathematical Foundations of Statistical Mechanics" (New York,N.Y., 1949)

(2) I.A. Ibragimov, Yu.V.Linnik, "Independent and Stationary Sequences of Random Variables" (Groningen, 1971)

(3) For the connection between the renormalization group and probability theory see
P.M.Bleher, Ya.G.Sinai, Comm.Math.Phys. 33, 23 (1973) and 45, 247 (1975)
G.Gallavotti, A.Martin-Löf, Nuovo Cimento 25B, 425 (1975)
M.Cassandro, G.Gallavotti, Nuovo Cimento 25B, 691 (1975)
G.Jona-Lasinio, Nuovo Cimento 26B, 99 (1975)
G.Gallavotti, G.Jona-Lasinio, Comm.Math.Phys. 41 301 (1975)
Ya.G. Sinai, Teor.Ver. i eë Prim. 21, 63 (1976)
R.L.Dobrushin, "Avtomodelnost i Renorm-gruppa Obobshchennykh Sluchainykh Polei", to be published (Communicated to us by B.Tirozzi)

(4) I.A.Ibragimov,Yu.A.Rozanov, "Processus Aléatoires Gaussiens" (Moscow, 1973) and Ref. (2)

(5) A.Martin-Löf, Comm.Math.Phys. 32, 75 (1973)
G.Gallavotti, A.Martin-Löf, Nuovo Cimento, 25B 425 (1975)

(6) G.Jona-Lasinio, in "Les méthodes mathématiques de la théorie quantique des champs", Colloques Internationaux C.N.R.S. N°. 248, pag 207 (Marseille 1975)

(7) R.L.Dobrushin, Theory of Prob. and its Appl. XIII, 197 (1968)

(8) See for example Ch. II of ref.(2)

(9) C.M.Newman, "Gaussian Correlation Inequalities for Ferromagnets", to appear in Z. für Wahrscheinlichkeitstheorie

(10) Yu.A. Davydov, Theory of Prob. and its Appl. XVIII, 312 (1973)

(11) G.C.Hegerfeldt, C.R.Nappi, "Mixing Properties in Lattice Systems", preprint 1976.

(12) B.M.McCoy, TT.Wu, "The two dimensional Ising model", (New York, 1973), Ch XI.

(13) C.M. Newman, Comm.Math.Phys. 41 1 (1975)

(14) E. Lukács, "Characteristic Functions" pg. 191 (London 1970)

SOLITONS AND BREATHERS

K. Pohlmeyer

Institute for Theoretical Physics

University of Heidelberg, Germany

I. Classical Soliton Physics

Let us discuss the notion "soliton" first in the context of one time- one space dimensional models. In a second step I shall explain how this notion can be generalized to three space dimensions [1].

For the sake of simplicity let us consider non-derivative self-couplings of neutral scalar particles of one sort in one time and one space dimension. Moreover, let us restrict ourselves to theories into which the parameter g playing the role of a coupling constant enters in the following specific way

$$L(t,x) = \tfrac{1}{2}\left(\frac{\partial\phi(t,x)}{\partial t}\right)^2 - \tfrac{1}{2}\left(\frac{\partial\phi(t,x)}{\partial x}\right)^2 - g^{-2}U(g\phi)$$

Here $L(t,x)$ is the Lagrangian density of the theory, $\phi(t,x)$ the scalar field, and $U(\xi)$ an infinite differentiable function which is assumed to have several global minima: $\xi = \xi_i$ $i = 1,2,\ldots$ and to adopt the value zero there: $U(\xi_i) = 0$, $i = 1,2,\ldots$ For instance, see Figure 1.

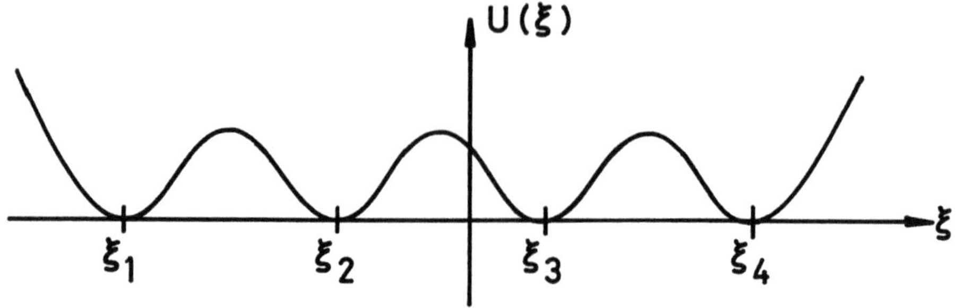

Figure 1

The velocity of light c is set equal to one.

The solutions of the Euler-Lagrange equation $\phi(t,x) \equiv \frac{\xi i}{g}$ which are constant in time and space correspond to the ground states of the system: degenerate vacua.

If we expand $U(g\phi)$ in a power series around one of the global minima and define

$$\varphi(t,x) = \phi(t,x) - \frac{\xi i}{g}$$

we find

$$L(t,x) = \frac{1}{2}\left(\frac{\partial\varphi(t,x)}{\partial t}\right)^2 - \frac{1}{2}\left(\frac{\partial\varphi(t,x)}{\partial x}\right)^2 - \frac{U''(\xi i)}{2!}\varphi^2(t,x) - g\,\frac{U'''(\xi i)}{3!}\varphi^3(t,x) + \ldots \,.$$

The term quadratic in φ is independent of g, the term cubic in φ is linearly dependent on g, etc.. Hence, the parameter g indeed plays the rôle of a coupling constant. Also, since $U(\xi)$ has a (global) minimum for $\xi = \xi_i$, $U''(\xi i)$ is larger than zero. As can be seen from the last formula $U''(\xi_i)$ is the squared mass μ_i^2 of the linearized field equation in the "i^{th} vacuum sector".

By the following redefinition of the field

$$\breve{u}(t,x) = g\phi(t,x)$$

the g-dependence can be completely removed from the equation of motion

$$\frac{\partial^2\breve{u}(t,x)}{\partial t^2} - \frac{\partial^2\breve{u}(t,x)}{\partial x^2} + U'(\breve{u}(t,x) = 0$$

and this - to a large extent - eliminates the g-dependence from the classical theory. However, the g-dependence cannot be eliminated from the fundamental equations of the corresponding quantum theory consisting of the operator equation of motion, the defining equation for the conjugate momentum and the equal time commutation relation. In other words, in the Feynman path integral not only the position of the extremal path matters but also the value of the action for the extremal path and the "higher expansion coefficients" of the action around the extremal path enter.

In a given reference frame the potential functional of our theory is

$$V[\phi] = \int_{-\infty}^{+\infty} dx \, \{\tfrac{1}{2}\phi_x^2 + g^{-2}\, U(g\phi)\}.$$

Indices x and t denote partial differentiations with respect to x and t respectively. We look for local minima $\phi(x) = \phi_c(x)$ of the potential energy functional describing classical equilibrium configurations of the system. The conditions are

1) $\left.\dfrac{\delta V[\phi]}{\delta\phi}\right/_{\phi=\phi_c} = 0 \; : \quad \dfrac{d^2\phi_c(x)}{dx^2} = g^{-1}.\, U'(g\phi_c(x))$

2) $0 < V[\phi_c] < +\infty$

3) $V[\phi_c] + \psi \geq V[\phi_c]$ for $\psi = \psi(x)$ sufficiently small in a certain topology, i.e.

$$\int_{-\infty}^{+\infty} dx\, \psi(x) \, \{-\frac{d^2}{dx^2} + U''(g\phi_c(x)\}\, \psi(x) \geq 0.$$

The equation

$$\frac{d^2\phi_c(x)}{dx^2} = g^{-1}\, U'(g\phi_c(x))$$

can be integrated once

$$\tfrac{1}{2}\left[\frac{d\phi_c(x)}{dx}\right]^2 = g^{-2} \cdot U(g\phi_c(x))$$

where the vanishing integration constant results from the second requirement, the finiteness of the potential energy for the equilibrium configuration.

Moreover, again the finiteness of the potential energy forces the solution $\phi_c(x)$ to approach a vacuum solution for large distances

$$\phi_c(x) \underset{x\to-\infty}{\longrightarrow} \frac{\xi_k}{g} , \quad \phi_c(x) \underset{x\to+\infty}{\longrightarrow} \frac{\xi_l}{g} .$$

The approach to the vacuum solution requires an infinite space interval. The derivative $\frac{d\phi_c(x)}{dx}$ can change its sign only at a point x such that $\phi_c(x)= \frac{\xi_i}{g}$ (compare the second last formula). Hence $\phi_c(x)$ never approaches for $x\to+\infty$ the same vacuum solution as for $x\to-\infty$: $\phi_c(x)$, the soliton-at-rest solution - as we shall call it - interpolates between adjacent vacua $\frac{\xi_i}{g}$ and $\frac{\xi_{i+1}}{g}$, and for every ordered pair of adjacent vacua $\frac{\xi_i}{g}$, $\frac{\xi_{i+1}}{g}$ there exists exactly one 1-parameter family of interpolating soliton-at-rest solutions of equal energy

$$\{\phi_c \ / \ \phi_c(x) = \phi_{i,i+1}(x-x_o),\ x_o \in \mathbb{R}^1 \},$$

$$\begin{aligned} E_{c\ell}[\phi_{i,i+1}(x-x_o)] &= \int_{-\infty}^{+\infty} dx \ \{\tfrac{1}{2}(\tfrac{d}{dx}\phi_{i,i+1}(x-x_o))^2 \\ &\quad + g^{-2}U(g\phi_{i,i+1}(x-x_o))\} \\ &= \int_{-\infty}^{+\infty} dx \ \{\tfrac{1}{2}(\tfrac{d}{dx}\phi_{i,i+1}(x))^2 \\ &\quad + g^{-2}U(g\phi_{i,i+1}(x)) \\ &= \int_{-\infty}^{+\infty} dx \ (\tfrac{d}{dx}\phi_{i,i+1}(x))^2 . \end{aligned}$$

To the ordered pair of adjacent vacua $\frac{\xi_{i+1}}{g}$, $\frac{\xi_i}{g}$ there corresponds exactly the 1-parameter family of soliton-at-rest solutions

$$\{\phi_c \ / \ \phi_c(x)= \phi_{i+1,i}(x-x_o)= \phi_{i,i+1}(x_o-x),\ x_o \in \mathbb{R}^1\}$$

of unchanged energy.

Schematically:

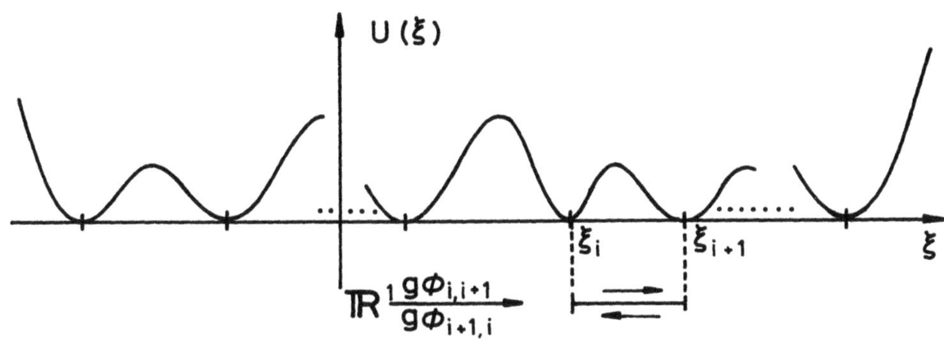

Figure 2

To recapitulate: There are some classical relativistic field theories which possess soliton-at-rest solutions, i.e. time-independent, but space-dependent stable solutions of the equation of motion the energy of which lies by finite amounts above the energy of the ground states. The latter ones are described by time- and space-independent solutions of the equation of motion. Solitons arise only when the ground state is degenerate.

We shall consider only those cases where the degeneracy is associated with a spontaneous breakdown of a symmetry: an accidental degeneracy of the classical ground state in general will be lifted in the corresponding quantum theory.

Let us investigate the stability of the soliton solutions in more detail. To this end, we consider solutions of the equation of motion whose Cauchy data at a given time are "close" to those of $\phi_{i,i+1}(x)$:

$$\phi(t,x) = \phi_{i,i+1}(x) + \psi(t,x)$$

As long as the "perturbation" remains small, $\psi(t,x)$ satisfies the linearized equation of motion

$$\{ \frac{\partial^2}{\partial t^2} - \frac{\partial^2}{\partial x^2} + U''(g\phi_{i,i+1}(x)) \} \psi(t,x) = 0.$$

Since this equation is linear and has time-independent coefficients, it suffices to consider "perturbations" $\psi(t,x)$ of the form $e^{i\omega t}\psi_\omega(x)$ where $\psi_\omega(x)$ satisfies the Schrödinger type eigenvalue equation

$$\{ - \frac{d^2}{dx^2} + U''(g\phi_{i,i+1}(x)) \}\psi_\omega(x) = \omega^2\psi_\omega(x) .$$

Stability, i.e. confinement of $\phi(t,x)$ to a "neighbourhood" of $\phi_{i,i+1}(x)$ for all times, requires that no eigenvalue of this equation is negative. This is exactly the condition 3), namely that the classical potential energy functional has a local minimum at $\phi(x)=\phi_{i,i+1}(x)$.

In order to discuss the potential function of the differential operator on the l.h.s. of the above eigenvalue equation, we note

$$U''(g\phi_{i,i+1}(x)) \xrightarrow[x\to-\infty]{} \mu_i^2, \quad U''(g\phi_{i,i+1}(x)) \xrightarrow[x\to+\infty]{} \mu_{i+1}^2 .$$

Since, however, $U(\xi)$ has a maximum between the two adjacent global minima ξ_i and ξ_{i+1}, $U''(g\phi_{i,i+1}(x))$ must take negative values in a finite region in space.

We shall assume - for the sake of simplicity and symmetry - that

$$\mu_i = \mu_{i+1} = \mu .$$

In the special typical examples which shall follow later this assumption will always be satisfied.

Then the potential function looks as follows

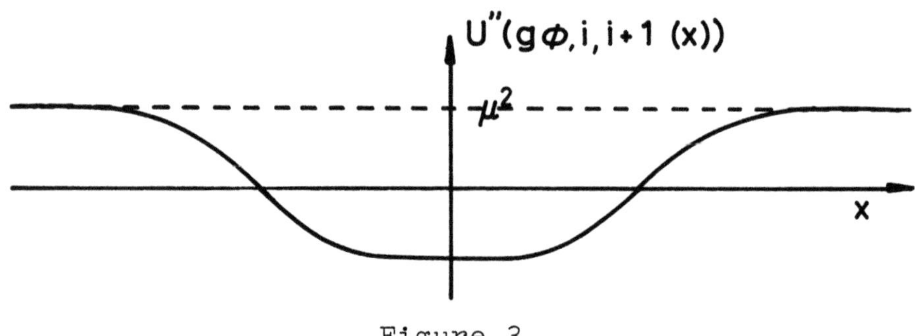

Figure 3

The differential operator

$$-\frac{d^2}{dx^2} + U''(g\phi_{i,i+1}(x))$$

has a continuous spectrum of eigenvalues

$$\omega^2 = \mu^2 + k^2$$

with

$$\psi^{\pm}_{\omega}(x) \approx e^{\pm ikx}, \quad \frac{d}{dx}\psi^{\pm}_{\omega}(x) \approx \pm ike^{\pm ikx}$$

for large negative distances and (possibly) some discrete bound states with $\omega^2<\mu^2$. We already know one such bound state corresponding to the eigenvalue $\omega^2 = 0$

$$\psi_o(x) = \frac{d}{dx}\phi_{i,i+1}(x)$$

because if we differentiate the equation

$$-\frac{d^2}{dx^2}\phi_{i,i+1}(x) + g^{-1}U'(g\phi_{i,i+1}(x)) = 0$$

once with respect to x, we obtain

$$\{-\frac{d^2}{dx^2} + U''(g\phi_{i,i+1}(x))\}\ (\frac{d}{dx}\phi_{i,i+1}(x)) = 0.$$

The existence of this eigenstate is a consequence of the translation invariance of the theory.

Stability requires that this state corresponds to the lowest eigenvalue. Hence, it must be the only eigenstate for the eigenvalue zero. In more than one space dimension the eigenvalue $\omega^2 = 0$ would be degenerate: the components of gradient $\phi_c(\underline{x})$ would be the corresponding eigenstates. Hence, for scalar fields without derivative coupling there would exist at least one negative eigenvalue which would destroy stability. To ensure stability in more than one space dimension one needs tensor forces, i.e. derivative coupling or spin which allow for a degenerate lowest eigenvalue.

The soliton solution is even stable against smooth local, but otherwise arbitrary perturbations of the dynamics. The following trivial conservation law is responsible for this stability

$$\partial^\mu j_\mu(t,x) = 0$$

$$j_\mu(t,x) = \epsilon_{\mu\nu}\partial^\nu\ \phi(t,x).$$

The charge density involves the spatial derivatives of the field only. Therefore, to this conservation law no symmetry of the theory corresponds. The conservation law and the associated conserved charge

$$Q = Q[\phi] = \int_{-\infty}^{+\infty} dx j_0(t,x)$$
$$= \phi(t,+\infty) - \phi(t,-\infty) = \frac{\xi_k}{g} - \frac{\xi_\ell}{g}$$

are of topological kind. The last equation holds for all finite energy solutions of the equation of motion since all of them must approach vacuum solutions for large distances. Continuity implies that $Q[\phi]$ remains unchanged under smooth perturbations of the dynamics as long as the perturbations are confined to bounded regions in space.

It is not possible to reach a solution with $k \neq 1$ by a perturbation expansion in g around the free field solution of any of the vacuum sectors. For $g = 0$, the solutions in question lie infinitely far apart from each other. The energy of the solution with $k \neq 1$ as well as its charge Q tend to infinity as g tends to zero, whereas the energy and the charge of the free field solution are independent of g. This divergence comes about by growing values of the field and not by an expansion of the region in which the energy is concentrated, i.e. the effective transition region from one vacuum solution to the other.

Granted that the energy difference between the soliton solutions and the vacuum solutions is finite for finite $g \neq 0$, both types of solutions are nevertheless separated from each other by an infinite potential barrier. For any continuous interpolating field function $\phi(x;\sigma)$ $-\infty < x < +\infty$, $0 < \sigma < 1$ with

$$\phi(x;0) = \frac{\xi_k}{g} \quad , \quad \phi(x;1) = \phi_{i,i+1}(x)$$

the potential energy function

$$V(\sigma) = V[\phi(x;\sigma)] = \int_{-\infty}^{+\infty} dx \{ \tfrac{1}{2} (\frac{d}{dx} \phi(x;\sigma))^2 + g^{-2} U(g\phi(x;\sigma)) \}$$

is infinite at least at one interior point of the interval $0 < \sigma < 1$. It is this infinite potential barrier which prevents the quantum mechanical soliton state from tunneling into states of a vacuum sector.

For every soliton-at-rest solution $\phi_{i,i+1}(x)$ Lorentz boosts provide new time-dependent solutions of the equation of motion

$$\phi_{i,i+1}\left(\frac{x - vt}{\sqrt{1 - v^2}}\right)$$

with energy

$$\frac{E_{c\ell}[\phi_{i,i+1}]}{\sqrt{1 - v^2}}$$

These solutions describe solitons moving with constant velocity v.
As a first typical example we choose ($\hbar = 1$) $U(\xi) = \frac{1}{2}(\xi^2 - m^2)^2$.

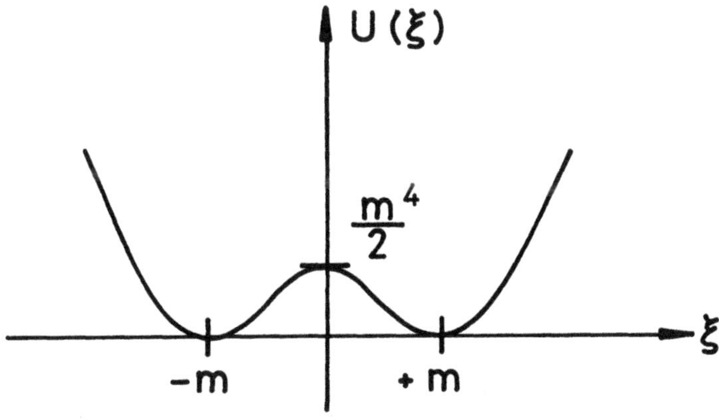

Figure 4

The Lagrangian is symmetric under the operation $\phi \leftrightarrow -\phi$, not so, however, the ground state solutions $\phi_i \equiv \frac{\xi_i}{g}$ $i = 1,2$, $\xi_1 = -m$, $\xi_2 = +m$: the discrete symmetry $\phi \leftrightarrow -\phi$ is spontaneously broken.
The masses μ_1 and μ_2 are both equal to 2m. The soliton-at-rest solution $\phi_{1,2}(x)$ with center of gravity in x = o is

$$\phi_{1,2}(x) = \frac{m}{g}\ \mathrm{tgh}\ (mx).$$

The corresponding classical energy is

$$E_{c\ell}[\phi_{1,2}] = \frac{4}{3}\ \frac{m^3}{g^2}\ .$$

The stability equation is

$$\{-\frac{d^2}{dx^2} + 4m^2 - \frac{6m^2}{\cos^2h(mx)}\}\psi_\omega(x) = \omega^2\psi_\omega(x).$$

This equation can be solved in closed form: the spectrum consists of a continuum $\omega^2=4m^2+k^2$ and the two discrete eigenvalues $\omega^2=3m^2$ and $\omega^2=0$. The corresponding unnormalized eigenfunctions are

$$\psi_{\pm\sqrt{4m^2+k^2}}(x) = e^{\pm ikx} \quad \left\{\begin{array}{l}\text{essentially the Jacobi po-}\\ \text{lynomial of degree 2 in}\\ \text{tgh(mx) with indices depen-}\\ \text{ding on k/m}\end{array}\right\}$$

$$\psi_{\sqrt{3}m}(x) = \frac{\sinh(mx)}{\cos^2h(mx)}$$

$$\psi_0(x) = \frac{1}{\cos^2h\,(mx)}$$

There are the following two alternatives for the interpretation of the soliton solutions $\phi_{1,2}(x)$ and $\phi_{2,1}(x)$:

1. $\phi_{1,2}(x)$ and $\phi_{2,1}(x)$ describe the same physical object. With this identification scattering solutions of an arbitrary number of these objects exist.
2. $\phi_{1,2}(x)$ and $\phi_{2,1}(x)$ describe different physical objects B and $\bar{B}$. In this case only those asymptotic scattering configurations of Bs and $\bar{B}$s are possible in which Bs and $\bar{B}$s occur by turns: the $\bar{B}$s screen the long range forces between the Bs and vice versa.

As a second example we choose the sine Gordon theory ($\hbar$ = 1).

$$U(\xi) = m^4(-\cos(\frac{\xi}{m})) = 2m^4\sin^2(\frac{\xi-j(2\pi m)}{2m})$$

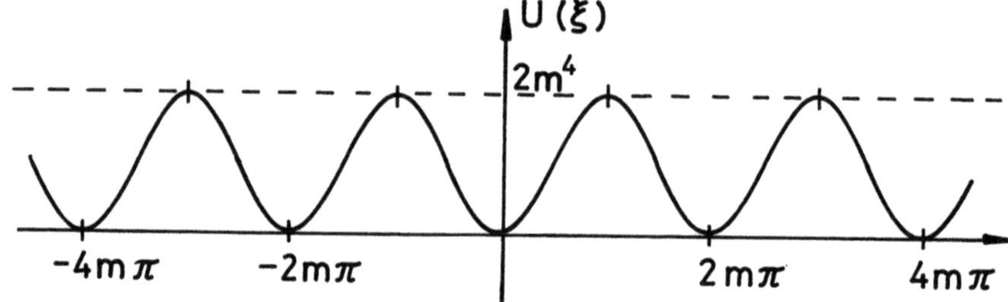

Figure 5

This theory has a very rich structure, a fact which was realized only a couple of years ago.

The Lagrangian is symmetric under the following discrete operations

$$\phi \leftrightarrow -\phi, \quad \phi \to \phi + k\left(\frac{2\pi m}{g}\right) \qquad k = 0, \pm 1, \pm 2, \ldots$$

These symmetries are not shared by the ground state solutions

$$\phi_i \equiv \frac{\xi_i}{g} \qquad i = 0, \pm 1, \pm 2, \ldots, \quad \xi_i = i(2\pi m)$$

The above mentioned discrete symmetries are spontaneously broken.
The masses μ_i are all equal to m. The soliton at rest solution $\phi_{i,i+1}(x)$ with center of gravity at x = o is

$$\phi_{i,i+1}(x) = \frac{4m}{g} \operatorname{arctg}(e^{mx}) + i\,\frac{2\pi m}{g}.$$

Its classical energy is

$$E_{c\ell}[\phi_{i,i+1}] = \frac{8m^3}{g^2},$$

its charge

$$Q = \frac{2\pi m}{g} \quad .$$

The solution

$$\phi_{i+1,i}(x) = \phi_{i,i+1}(-x)$$

of equal energy and opposite charge is called antisoliton-at-rest solution.
Solutions which differ by a constant additive multiple of $\frac{2\pi m}{g}$ only are identified.
The stability equation of the soliton-at-rest solution is

$$\left\{-\frac{d^2}{dx^2} + m^2 - \frac{2m^2}{\cos^2 h(mx)}\right\} \psi_\omega(x) = \omega^2 \psi_\omega(x).$$

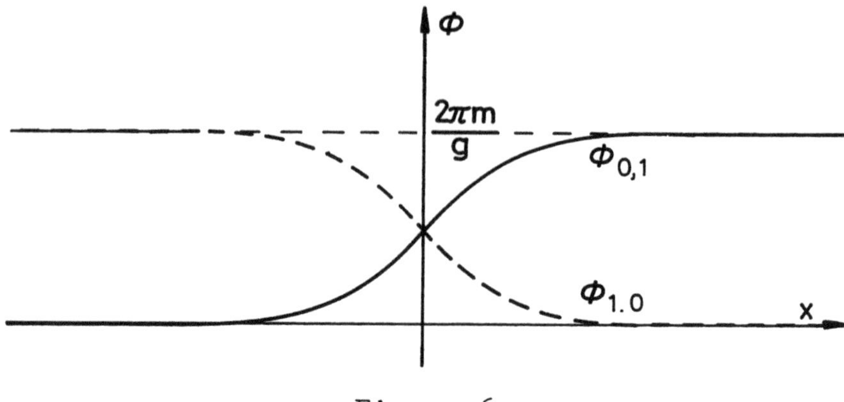

Figure 6

Again, this equation can be solved in closed form. The spectrum consists of a continuum $\omega^2 = m^2 + k^2$ and the discrete eigenvalue $\omega^2 = 0$. The corresponding unnormalized eigenfunctions are

$$\psi^{\pm}_{\sqrt{m^2+k^2}}(x) = e^{\pm ikx} \cdot \left\{ \begin{array}{l} \text{essentially the Jacobi} \\ \text{polynomial of degree 1 in} \\ \text{tgh(mx) with indices de-} \\ \text{pending on k/m,} \end{array} \right\}$$

$$\psi_o(x) = \frac{1}{\cosh(mx)} \quad .$$

There are scattering solutions which in the remote past and distant future can be described as a superposition of n_+ solitons and n_- antisolitons, where n_+ and n_- are arbitrary, but fixed, non-negative integers. The charge acts additively on these asymptotic configurations

$$Q = \frac{2\pi m}{g} (n_+ - n_-) \; .$$

Analytic expressions are available for these (n_+) soliton-(n_-) antisoliton scattering solutions. For instance, the soliton-antisoliton scattering solution with its center of gravity at rest in x = o and with velocity u of the antisoliton relative to the center of gravity is given by

$$\phi_{SA}(t,x;-u,u,-\frac{u\Delta}{2},\frac{u\Delta}{2}) = \frac{4m}{g}\operatorname{arctg}\left\{\frac{\sinh(t\frac{mu}{\sqrt{1-u^2}})}{u\cosh(x\frac{m}{\sqrt{1-u^2}})}\right\}$$

where $\Delta = 2\frac{\sqrt{1-u^2}}{mu}\ln u<0$ and where the arguments of ϕ_{SA} denote in turn time and space coordinate, soliton and antisoliton velocity relative to the center of gravity, and the position of the center of the soliton and antisoliton at time $t = o$ when "freely" extrapolated from time $t = -\infty$.

In the remote past

$$\phi_{SA}(t,x;\ldots) \underset{t\to-\infty}{\longrightarrow} \phi_{o,-1}\left(\frac{x-u(t+\frac{\Delta}{2})}{\sqrt{1-u^2}}\right) + \phi_{-1,o}\left(\frac{x+u(t+\frac{\Delta}{2})}{\sqrt{1-u^2}}\right)$$

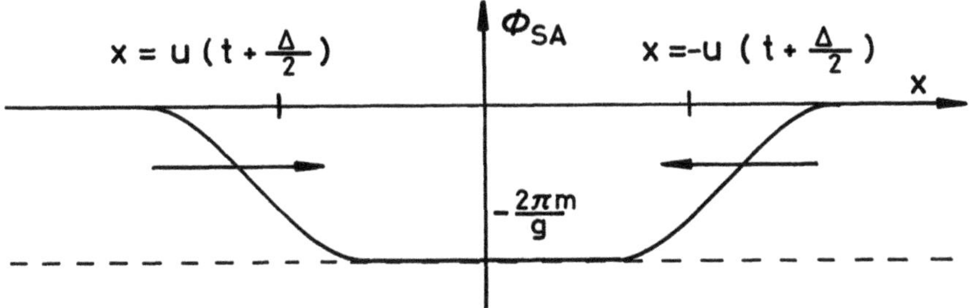

Figure 7. Antisoliton, $x = u(t+\frac{\Delta}{2})$; Soliton, $x = -u(t+\frac{\Delta}{2})$.

In the distant future

$$\phi_{SA}(t,x;\ldots) \underset{t\to+\infty}{\longrightarrow} \phi_{o,1}\left(\frac{x+u(t-\frac{\Delta}{2})}{\sqrt{1-u^2}}\right) + \phi_{1,o}\left(\frac{x-u(t-\frac{\Delta}{2})}{\sqrt{1-u^2}}\right)$$

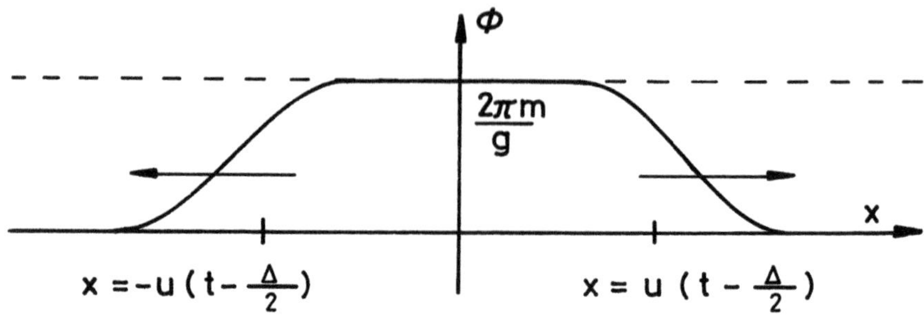

Figure 8. Soliton, $x = -u(t-\frac{\Delta}{2})$; Antisoliton, $x = u(t-\frac{\Delta}{2})$.

In all such scattering processes the number of solitons and the number of antisolitons are separately conserved. Moreover, even the momenta of the incoming solitons coincide with the momenta of the outgoing solitons. The same is true for the antisolitons. Classically solitons and antisolitons attract and penetrate each other, solitons repel and reflect each other, and so do the antisolitons. The classical scattering can be viewed as a series of subsequent two-body collisions [2].

Apart from the scattering solutions there exist also periodic solutions of finite energy which can be interpreted as bound states of a soliton-antisoliton pair. The binding energy takes values from a continuous range. Also these solutions are available in closed form. For instance, the "breather" solution at rest with its center of gravity at x = o and with period τ is given by

$$\phi_\tau(t,x) = \frac{4m}{g} \operatorname{arctg} \left(\sqrt{\tilde{\tau}^2-1} \cdot \frac{\sin\left(m \frac{t-t_o}{\tilde{\tau}}\right)}{\cosh\left(mx\frac{\sqrt{\tilde{\tau}^2-1}}{\tilde{\tau}}\right)} \right) .$$

Its energy is

$$E_{c\ell}[\phi_\tau] = \frac{16m^3}{g^2} \frac{\sqrt{\tilde{\tau}^2-1}}{\tilde{\tau}}$$

Here

$$\tilde{\tau} = \frac{m\tau}{2\pi} .$$

After so many analytic results it is no longer totally surprising that Fadeev and Takhtadzhyan could prove the following fact: The sine Gordon theory defines a completely integrable Hamiltonian system. In the second lecture I shall sketch the proof of this remarkable fact and describe the canonical transformation which effects the separation of variables for the Hamilton-Jacobi equations.

Originally [3] the name "soliton" was coined for solitary waves which emerge unchanged from scattering at each other, as for instance the solitary waves of the sine Gordon equation. At the bottom of this extreme stability there is a denumerably infinite set of local covariant (explicitly known) conservation laws. In the meantime, however, the name "soliton" has been used in a much broader sense for solutions, for which

$$\lim_{t\to\infty} \max_{\underline{x}} \Theta_{oo}(t,\underline{x}) \neq 0$$

where $\Theta_{oo}(t,\underline{x})$ denotes the energy density at time t at the point $\underline{x}$.

The concept of solitons is not restricted to one space dimension. As a last topic of this first lecture we shall discuss a relativistic local theory in three space dimensions which describes the interaction of an isotriplet of scalar Higgs fields ϕ_a and a triplet of gauge fields A^μ_a, whose equations of motion admit static (stable?) solutions of finite energy [4] . The Lagrangian density of the theory is

$$\mathscr{L} = -\frac{1}{4} G^{\mu\nu}_a G_{a\mu\nu} + \frac{1}{2}(D_\mu\phi)_a(D^\mu\phi)_a - g^{-2}U(g\phi)$$

where

$$G^{\mu\nu}_a = \partial^\mu A^\nu_a - \partial^\nu A^\mu_a + e\epsilon_{abc}A^\mu_b A^\nu_c$$

$$(D^\mu\phi)_a = \partial^\mu\phi_a + e\epsilon_{abc} A^\mu_b \phi_c$$

$$\phi = (\phi_a\phi_a)^{\frac{1}{2}}$$

$$U(\xi) = \frac{m^4}{2} [1 - \frac{\xi^2}{m^2}]^2 \quad .$$

The symbol ϵ_{abc} stands for the totally antisymmetric tensor in three dimensions. The Lagrangian is invariant under local SO_3 gauge transformations. The vacuum solutions

$$A^\mu_a \equiv 0 \quad a = 1,2,3 \quad \mu = 0,1,2,3; \; \phi_a \equiv n_a \frac{m}{g} \quad a = 1,2,3$$

spontaneously break this symmetry. Here n_a denotes a constant unit vector in three dimensional space.
With the help of the direction field

$$\hat{\phi}_a(x) = \phi_a(x)/\phi(x) \quad x = (x^0,..,x^3) = (t,\underline{x})$$

we define the gauge invariant tensor field

$$F^{\mu\nu} = \hat{\phi}_a \, G^{\mu\nu}_a - e^{-1} \epsilon_{abc} \, \hat{\phi}_a (D^\mu \hat{\phi})_b (D^\nu \hat{\phi})_c$$

the analogue to the electromagnetic field tensor.
$F^{\mu\nu}$ may be written in the form

$$F^{\mu\nu} = \{\partial^\mu(\hat{\phi}_a \, A^\nu_a) - \partial^\nu(\hat{\phi}_a \, A^\mu_a)\} + \{e^{-1}\epsilon_{abc}\hat{\phi}_a(\partial^\mu\hat{\phi}_b)(\partial^\nu\hat{\phi}_c)\}.$$

Note that the first bracket does not contribute to

$$k_\mu = \tfrac{1}{2}\epsilon_{\mu\nu\rho\sigma}\partial^\nu F^{\rho\sigma}$$

the symbol $\epsilon_{\mu\nu\rho\sigma}$ denoting the totally antisymmetric tensor in four dimensions.

Note further that the second bracket contains only Higgs fields and usual derivatives

$$k_\mu = \frac{1}{2e} \epsilon_{\mu\nu\rho\sigma} \epsilon_{abc} \, \partial^\nu[\hat{\phi}_a(\partial^\rho\hat{\phi}_b) \, (\partial^\sigma\hat{\phi}_c)]$$

$$\partial^\mu k_\mu = 0.$$

The "magnetic charge" M

$$M = \frac{1}{4\pi} \int d^3\underline{x} \; k_0(t,\underline{x})$$

$$= \frac{1}{8\pi e} \int d^3\underline{x} \; \epsilon_{ijk} \, \epsilon_{abc} \; \partial^i \, [\hat{\phi}_a(\partial^j\hat{\phi}_b) \, (\partial^k\hat{\phi}_c)]$$

does not define the infinitesimal generator of a continuous symmetry group. The charge density k_0 does not contain the canonically conjugate momenta. M is a topological charge.

We evaluate the integral with the help of the theorem by Gauss and Ostrogradski:

$$M = \lim_{R\to\infty} \frac{1}{4\pi e} \int_{S_R^2} (d^2\sigma)_i \frac{1}{2} \epsilon_{ijk} \epsilon_{abc} [\hat{\phi}_a (\partial^j \hat{\phi}_b)(\partial^k \hat{\phi}_c)]$$

Here we integrate over the surface of the infinitely remote sphere $|\underline{x}| = R$. The direction field ϕ_a provides a continuous mapping from $S_R^2 \to S_1^2$, the surface of the unit sphere in three-dimensional isotopic spin space. The set of these mappings has infinitely many connectivity components. The various components are characterized by the "degree of mapping" d which is defined as follows: If all of S_R^2 is mapped into a single point of S_1^2, then $d = 0$. Otherwise one picks a regular image point of S_1^2. Its finitely many originals on S_R are mapped into it d_+ times with positive and d_- times with negative orientation. The integer $d = d_+ - d_-$ is independent of the choice of the regular image point and is called the degree of mapping. It is a topological invariant.

If in the last expression for M we parametrize the surface S_R^2 by coordinates ξ^1, ξ^2, then the square of the integrand is equal to the determinant of the metric tensor

$$\left(\left(\frac{\partial}{\partial\xi^\alpha} \hat{\phi}_a\right) \left(\frac{\partial}{\partial\xi^\beta} \hat{\phi}_a\right) \right)$$

of the corresponding multifarious parametrization of the surface S_1^2. Since S_1^2 is covered d_+ times with positive and d_- times with negative orientation, the integral yields $d4\pi$. Hence $M = \frac{d}{e}$.

It remains to be shown that there exist static solutions $\phi_a(x)$, $A_a(x)$ of finite energy with $d \neq 0$. To construct such a solution with $d = 1$ we write down the following Ansatz due to 't Hooft:

$$\phi_a(t,x) = \phi_a(\underline{x}) = \frac{x^a}{r} F(r)$$

$$A_a^0(t,\underline{x}) \equiv 0, \quad A_a^i(t,\underline{x}) = A_a^i(\underline{x}) = e^{-1} \epsilon_{aij} \frac{x^j}{r^2} A(r)$$

with

$$r = |\underline{x}| .$$

$\hat{\phi}_a(\underline{x}) = \frac{x^a}{r}$ maps the points of S_R^2 into the points "below" them of S_1^2:

$$d = 1 \ .$$

The finiteness of the energy is ensured by a sufficiently rapid convergence of $F(r) \underset{r\to\infty}{\to} \frac{m}{g}$, $A(r) \underset{r\to\infty}{\to} 1$. The differential equations for F and A (derived from the equation of motion) possess smooth solutions behaving for r tending to infinity as indicated, and for r tending to zero as

$$F(r) = O(r), \quad A(r) = O(r^2) \ .$$

Thereby it is proven that the model under consideration has a static solution of finite energy and magnetic charge e^{-1}, the 't Hooft magnetic monopole.

For other local gauge theories with a spontaneously broken symmetry group $\mathcal{G}$ and degenerate vacua, all of which can be reached from a particular vacuum by applying symmetry transformations, there will be non-trivial topological charges, provided the continuous mappings from

$$S_R \longrightarrow \mathcal{G}/\mathcal{H}$$

consist of several homotopy classes. Here $\mathcal{G}/\mathcal{H}$ is the set of all conjugate subgroups $g\mathcal{H}g^{-1}$ with $g \in \mathcal{G}$ and the unbroken subgroup $\mathcal{H}$, the stabilizer of the particular vacuum. These conjugate subgroups are in one-to-one corresponding to the degenerate vacua. In four and more space dimensions solitons do not exist which are associated with mappings of non-trivial degree (provided by the matter fields) of the infinitely remote sphere to the set of vacua [5]. For other possibilities compare ref.[6].

II. The Sine Gordon Theory as a Completely Integrable Hamiltonian System

In the two remaining lectures we shall be concerned exclusively with the sine Gordon theory.

It has already been mentioned that the g dependence can be eliminated from the equation of motion

$$\frac{\partial^2}{\partial t^2}\phi - \frac{\partial^2}{\partial x^2}\phi + g^{-1}m^3 \sin\left(\frac{g}{m}\phi\right) = 0$$

by rescaling the field ϕ by a factor proportional to g. Actually, by passing to the dimensionless coordinates and fields

$$x' = mx, \quad t' = mt, \quad \breve{u}(t',x') = \frac{g}{m}\phi(t,x)$$

the g and m dependence simultaneously can be removed from the equation of motion. In the new coordinates the Lagrangian density reads

$$\mathcal{L}(t,x) = \mathcal{L}'(t',x') = \frac{1}{2\gamma}\{\breve{u}_{t'}(t',x')^2 - \breve{u}_{x'}(t',x')^2 + 2(\cos(\breve{u}(t',x')) - 1)\}$$

where

$$\gamma = \frac{g^2}{m^4} \quad .$$

Once and forever we treat the theory in the new coordinates and drop primes:

$$\mathcal{L} = \frac{1}{2\gamma}\{\breve{u}_t^2 - \breve{u}_x^2 + 2(\cos\breve{u} - 1)\}$$

$$\left(\frac{\partial^2}{\partial t^2} - \frac{\partial^2}{\partial x^2}\right)\breve{u} + \sin\breve{u} = 0 \ .$$

Next, we pass to the dimensionless light-cone coordinates

$$\xi = \frac{t+x}{2} \ , \ \eta = \frac{t-x}{2}$$

in which the d'Alembertian factorizes

$$\Box = \frac{\partial^2}{\partial t^2} - \frac{\partial^2}{\partial x^2} = \frac{\partial^2}{\partial\xi\partial\eta}$$

and in which the equation of motion takes the simple form

$$\breve{u}_{\xi\eta} + \sin\breve{u} = 0 \ .$$

The equation of motion is invariant under Lorentz transformations $L_\zeta, \zeta \in \mathbb{R}^1 - \{0\}$

$$L_\zeta \ : \ \xi,\eta \to \zeta\xi, \zeta^{-1}\eta$$

$$u(\xi,\eta) \to \breve{u}^{(\zeta)}(\xi,\eta) = \breve{u}(\zeta^{-1}\xi, \zeta\cdot\eta)$$

II. 1. Derivation of the Isospectral Family of Associated Linear Eigenvalue Problems

There exists a non-linear Bäcklund transformation

T_1 of the set of solutions of the equation of motion into itself defined (modulo boundary conditions) by the two ordinary differential equations

$$T_1 \begin{cases} \left(\frac{\breve{u}(.;1)+\breve{u}}{2}\right)_\xi = \sin\left(\frac{\breve{u}(.;1)-\breve{u}}{2}\right) \\ \left(\frac{\breve{u}(.;1)-\breve{u}}{2}\right)_\eta = -\sin\left(\frac{\breve{u}(.;1)+\breve{u}}{2}\right) \end{cases}$$

with the consistency requirement

$$\breve{u}_{\xi\eta}(.;1) + \sin\breve{u}(.;1) = 0, \quad \breve{u}_{\xi\eta} + \sin\breve{u} = 0 .$$

The Bäcklund transformation T_1 does not commute with the Lorentz transformations L_ζ. Hence by forming

$$T_\zeta = L_\zeta T_1 L_\zeta^{-1} \begin{cases} \left(\frac{\breve{u}(.;\zeta)+\breve{u}}{2}\right)_\xi = \zeta^{-1}\sin\left(\frac{\breve{u}(.;\zeta)-\breve{u}}{2}\right) \\ \left(\frac{\breve{u}(.;\zeta)-\breve{u}}{2}\right)_\eta = -\zeta\sin\left(\frac{\breve{u}(.;\zeta)+\breve{u}}{2}\right) \end{cases}$$

we obtain a whole one-parameter family of Bäcklund transformations. Now we would like to derive the isospectral family of linear eigenvalue problems in light-cône coordinates

$$L(\eta)\psi = \varkappa\psi$$

which plays a central role in the inverse scattering method. The η-coordinate is the deformation parameter; it labels the members of the isospectral family.

We start from the first of the two differential equations defining T_ζ

$$\left(\frac{\breve{u}(.;\zeta)+\breve{u}}{2}\right)_\xi = \zeta^{-1}\sin\left(\frac{\breve{u}(.;\zeta)-\breve{u}}{2}\right)$$

and reduce its transcendental non-linearity to a quadratic one by substituting for $u(.;\zeta)$

$$\Gamma = \mathrm{tg}\left(\frac{\bar{u}(.;\zeta)-\bar{u}}{4}\right) \quad :$$

$$\Gamma_\xi + \frac{\bar{u}_\xi}{2}(1+\Gamma^2) = \zeta^{-1}\Gamma .$$

This is a one-parameter family of Riccati equations. These can be linearized by the following ansatz

$$\Gamma = \frac{\psi_1}{\psi_2} \quad .$$

The resulting differential equation

$$\psi_2\,\psi_{1\xi} - \psi_1\,\psi_{2\xi} + \frac{u_\xi}{2}(\psi_1^2 + \psi_2^2) = \zeta^{-1}\psi_1\psi_2$$

is satisfied if ψ_1 and ψ_2 solve the following linear system of first order differential equations

$$\psi_{1\xi} + \frac{\breve{u}_\xi}{2}\psi_2 = (2\zeta)^{-1}\psi_1$$

$$-\psi_{2\xi} + \frac{u_\xi}{2}\psi_1 = (2\zeta)^{-1}\psi_2 \quad .$$

This is a linear eigenvalue problem for each value of η

$$L(\eta)\psi = \varkappa\,\psi$$

where

$$\psi = \begin{pmatrix}\psi_1\\ \psi_2\end{pmatrix}$$

$$L(\eta) = i\begin{pmatrix}1 & 0\\ 0 & -1\end{pmatrix}\frac{d}{d\xi} + \frac{i}{2}\bar{u}_\xi\begin{pmatrix}0 & 1\\ 1 & 0\end{pmatrix}$$

$$\varkappa = \frac{i}{2\zeta} \; .$$

The η-evolution of ψ can be determined from the second of the two ordinary differential equations defining T_ζ :

$$\frac{\partial\psi}{\partial\eta} = -\,M\psi$$

with

$$M = -\frac{1}{4i\varkappa}\begin{pmatrix}\cos\breve{u}, & \sin\breve{u}\\ \sin\breve{u}, & -\cos\breve{u}\end{pmatrix} .$$

We confirm the relation

$$\frac{\partial L}{\partial \eta} = [L,M]$$

which guarantees that the spectra of differential operators $L(\eta)$ for different values of η coincide. To see this, we differentiate the equation

$$L(\eta)\psi = \varkappa\psi$$

once with respect to η and use the last relation. Thus, we find

$$(\frac{\partial\varkappa}{\partial\eta})\psi = 0$$

whence

$$\frac{\partial\varkappa}{\partial\eta} = 0 .$$

In particular, the discrete spectrum $\varkappa_1,\ldots,\varkappa_N$ does not depend on η:

$$N(\eta) = N(0) = N, \quad \varkappa_j(\eta) = \varkappa_j(0) = \varkappa_j \quad j=1,\ldots,N .$$

II.2. The Inverse Scattering Method and the Characteristic Initial Value Problem

Why do we care about this isospectral family of associated linear eigenvalue problems? We care about it because with its help - in the best tradition of physics - an interesting non-linear problem, namely the η-evolution of a general finite energy solution of the characteristic initial value problem, can be reduced to a chain of linear problems. Without loss of generality the finite energy characteristic initial value problem is specified by giving $u_\xi(\xi,0)\; -\infty<\xi<+\infty$ ($\lim_{\xi\to-\infty} u(\xi,\eta)=0$ for $\forall\,\eta \geq 0$).

The "linear" chain consists of the following steps:

1. The characteristic initial data $u_\xi(\xi,0)\; -\infty<\xi<+\infty$ fix the potential of a linear differential operator $L(0)$ belonging to an isospectral family.
2. The scattering data of the operator $L(0)$ are computed

(direct scattering problem). They consist of the reflection coefficient and the discrete eigenvalues together with complex numbers which for the corresponding bound state wave functions determine the ratio of the coefficients in front of the exponential damping factors for $\xi\to+\infty$ and $\xi\to-\infty$: $s(0)=\{r(\lambda,0)-\infty<\lambda+\infty;\ \varkappa_j, m_j(0)\ j=1,..,N\}$.

3. The equations for the η-evolution of the scattering data are integrated explicitly using the isospectrality of the operators $L(\eta)$ and the finiteness of the energy.

4. The potential of the linear differential operator $L(\eta)$ is obtained from the scattering data

   $$s(\eta) = \{r(\lambda,\eta)-\infty<\lambda<+\infty;\ \varkappa_j,\ m_j(\eta) \quad j = 1,...,N\}$$

   by means of the Gelfand-Levitan-Marchenko equation (inverse scattering problem).

5. Finally, the characteristic initial data $\mathring{u}_\xi(\xi,\eta)-\infty<\xi<+\infty$ are extracted from the potential of the operator $L(\eta)$.

The linear differential operators need not be self-adjoint.

Let me now remind you how to solve

a) the direct scattering problem
b) the inverse scattering problem
for self-adjoint second order differential operators on the entire real line [7].

a) Determine the scattering data of the operator

$$L = L_o + V(x) \quad -\infty < x < +\infty\ ,\ L_o = -\frac{d^2}{dx^2}$$

with the real Lebesgue integrable function $V(x)$ satisfying

$$\int_{-\infty}^{+\infty} dx\ (1+|x|)|V(x)| < \infty!$$

Let $f_1(x,k)$ and $f_2(x,k) = f_1(x,-k)$ be the fundamental solutions normalized at $+\infty$ of the eigenvalue equation

$$L\psi = k^2\psi$$

and $g_1(x,k)=g_2(x,-k)$, $g_2(x,k)$ the fundamental solutions normalized at $-\infty$, e.g.

$$g_2(x,k) \underset{x\to-\infty}{\approx} e^{-ikx} \quad , \quad \frac{d}{dx} g_2(x,k) \underset{x\to-\infty}{\approx} -ike^{-ikx} \quad .$$

The differential equation for $g_2(x,k)$ plus the Sommerfeld radiation condition can be converted into the integral equation of Volterra type

$$g_2(x,k) = e_2(x,k) + \int_{-\infty}^{x} dx' \, \frac{\sin k(x-x')}{k} \, V(x') g_2(x',k) \quad .$$

Here $e_1(x,k)=e^{ikx}$ and $e_2(x,k)=e^{-ikx}$ are the fundamental solutions of the free eigenvalue equation

$$L_o\psi = k^2\psi \quad .$$

Set

$$g_2(x,k) - e_2(x,k) = h_2(x,k) \quad .$$

$h_2(x,k)$ is analytic in the variable k in the upper half k-plane and continuous in the closed upper half k-plane. Further, for $|k|\to\infty$, $\mathrm{Im}k > 0$ and fixed real x

$$h_2(x,k)\cdot e_2(x,k)^* = O\left(\frac{1}{|k|}\right) \quad .$$

It then follows from the Paley-Wiener theorem that g_2 can be represented in the following form

$$g_2(x,k) = e_2(x,k) + \int_{-\infty}^{x} dy \, B_2(x,y)\cdot e_2(y,k)$$

where $B_2(x,y) \in L_2(\mathbb{R}_y)$ for every fixed real x and where $B_2(x,y)$ satisfies the following integral equation

$$B_2(x,x+2y) = \tfrac{1}{2}\int_{-\infty}^{x+y} dt \, V(t) + \int_{y}^{o} dz \int_{\infty}^{x+y-z} dt V(t) B_2(t,t+2z)$$

for $y < 0$.

$B_2(x,x)$ is related to $V(x)$ through the equation

$$2 \frac{d}{dx} B_2(x,x) = V(x) \quad .$$

The fundamental solutions $f_1(x,k)$, $f_1(x,-k)$ can be expressed as linear combinations of the fundamental solutions $g_2(x,-k)$, $g_2(x,k)$ and vice versa. For instance

$$f_1(x,k) = a(k)\, g_2(x,-k) + b(k)\, g_2(x,k)$$

$$g_2(x,k) = -b(-k)\, f_1(x,k) + a(k)\, f_1(x,-k)$$

The coefficients a(k) and b(k) have the following representations in terms of $B_2(x,y)$

$$a(k) = 1 - \frac{1}{2ik}\int_{-\infty}^{+\infty} dt\, V(t) - \frac{1}{2ik}\int_{0}^{\infty} ds\, e^{2iks}\, \Pi_2(s)$$

$$b(k) = \frac{1}{2ik}\int_{-\infty}^{+\infty} ds\, e^{2iks}\, \Pi_1(s)$$

where

$$\Pi_1(s) = V(s) + 2\int_{s}^{\infty} dt\, V(t)\, B_2(t,2s-t)$$

$$\Pi_2(s) = 2\int_{-\infty}^{+\infty} dt\, V(t)\, B_2(t,t-2s)\ .$$

$a(\lambda)$ is analytic (continuous) in the (closed) upper half k-plane. It has finitely many zeros there. All of them lie on the imaginary axis and are simple. They correspond to the discrete eigenvalues $k = i\varkappa_j$ $j = 1,\ldots,N$ of the operator L.

For $k = i\varkappa_j$

$$f_1(x,k\varkappa_j) = b_j\, g_2(x,i\varkappa_j)\ .$$

In order to solve the direct scattering problem determine $B_2(x,y)$ from the above integral equation and compute b(k) and a(k) from the above formulae. The scattering data are then given by

$$r(k) = \frac{b(k)}{a(k)} \quad -\infty < k < +\infty;\quad i\varkappa_j,\ m_j = \frac{b_j}{i\frac{d}{dk}a(k)_{/k=i\varkappa_j}} \quad j=1,\ldots,N.$$

b) Determine the potential V(x) of the operator

$$L = L_o + V(x) \qquad -\infty < x < +\infty$$

from the scattering data

$$\{r(\lambda) \quad -\infty < \lambda < +\infty; \quad i\varkappa_j, \ m_j \quad j = 1,\dots,N\}$$

which are assumed to have property 6) of theorem 3.4 of ref.[7] !

First, we express a(k) in terms of the scattering data

$$a(k) = \lim_{\varepsilon \downarrow o} \ a(k+i\varepsilon) = \lim_{\varepsilon \downarrow o} \left\{ \exp\left(\frac{1}{2\pi i}\int_{\infty}^{\infty} dk' \ \frac{\ell n(1-|r(k')|^2)}{-k'+(k+i\varepsilon)}\right) \prod_{1}^{N} \left(\frac{(k+i\varepsilon)-i\varkappa_j}{(k+i\varepsilon)+i\varkappa_j}\right)\right\} .$$

Next we write down a slight modification of the equation defining the coefficients a(k) and b(k) = r(k) · a(k)

$$\left[\frac{1}{a(k)} - 1\right] f_1(x,k) = r(k)h_2(x,k) + r(k)e_2(x,k)+h_2(x,-k) - [f_1(x,k) - e_1(x,k)] .$$

We multiply this equation by $\frac{e_2(y,k)}{2\pi}$, integrate both sides for x > y over the entire real axis, evaluate the l.h.s. by the residue theorem and insert the integral representation for $h_2(x,k) = g_2(x,k) - e_2(x,k)$. In this way we obtain the linear Gelfand-Levitan-Marchenko integral equation for $B_2(x,y)$:

$$x > y \ : \ B_2(x,y)+\mathcal{F}(x,y)+\int_{-\infty}^{x} du \ B_2(x,u)\mathcal{F}(u,y) = 0$$

where the kernel $\mathcal{F}(x,y)$ is determined by the scattering data:

$$\mathcal{F}(x,y) = \frac{1}{2\pi}\int_{-\infty}^{+\infty} dk \ r(k) \ e^{-ik(x+y)} + \sum_{j=1}^{N} m_j \ e^{\varkappa_j(x+y)} .$$

We compute $B_2(x,y)$ from the integral equation and find the potential $V(x)$ from the relation

$$V(x) = 2 \frac{d}{dx} B_2(x,x) .$$

Let us now return to our isospectral family of linear eigenvalue problems

$$L(\eta)\psi = \varkappa\psi \quad .$$

We want to show that it is a simple matter to establish the η-evolution of the scattering data.

We start from the equation

$$\frac{\partial\psi(\xi;\varkappa,\eta)}{\partial\eta} = - M\psi(\xi;\varkappa,\eta) \quad .$$

Since $\tilde{u}(\xi,\eta)$ is not known, also M is not known. However, the η-evolution of the scattering data involves only the asymptotic form of M. We are interested in solutions with finite energy (and momentum). Hence $\cos \tilde{u}(\xi,\eta) \underset{|\xi|\to\infty}{\to} 1$ for all η. Thus, for the solutions of interest the asymptotic form of the operator M is known

$$M \approx - \frac{1}{4i\varkappa} \begin{pmatrix} 1 & 0 \\ 0 & -1 \end{pmatrix} \quad .$$

Inserting the asymptotic forms of M and ψ

$$\psi(\xi;\varkappa,\eta) \underset{\xi\to-\infty}{\approx} a(\varkappa,\eta)e^{i\varkappa\xi} \begin{pmatrix} 0 \\ 1 \end{pmatrix} + b(\varkappa,\eta)e^{-i\varkappa\xi}\begin{pmatrix} 1 \\ 0 \end{pmatrix}$$

$$\underset{\xi\to+\infty}{\approx} e^{i\varkappa\xi} \begin{pmatrix} 0 \\ 1 \end{pmatrix}$$

$$\psi(\xi;i\varkappa_j,\eta) \underset{\xi\to-\infty}{\approx} b_j(\eta)e^{-i\varkappa_j\xi} \begin{pmatrix} 1 \\ 0 \end{pmatrix}$$

$$\underset{\xi\to+\infty}{\approx} e^{i\varkappa_j\xi} \begin{pmatrix} 0 \\ 1 \end{pmatrix}$$

into the above equation and taking the normalization for ψ into account, we arrive at

$$a(\varkappa,\eta)=a(\varkappa,\ o),\ b(\varkappa,\eta) = \exp\{\frac{\eta}{2i\varkappa}\}\ b(\varkappa,\ o),$$

$$b_j(\eta) = \exp\{\frac{\eta}{2i\varkappa_j}\}\ b_j(o)$$

i.e.

$$r(\varkappa,\eta) = \exp\{\frac{\eta}{2i\varkappa}\}\ r(\varkappa,o),\ \varkappa_j(\eta) = \varkappa_j,$$

$$m_j(\eta) = \exp\{\frac{\eta}{2i\varkappa_j}\}\ m_j(o).$$

Multi-soliton, multi-antisoliton scattering solutions correspond to reflectionless potentials: $r(\varkappa,\eta) \equiv 0$ with

$$a(\varkappa,\eta)=a(\varkappa,o)= \prod_{j=1}^{N} \left(\frac{\varkappa - \varkappa_j}{\varkappa - \varkappa_j^*}\right) \qquad \varkappa_j \text{ purely imaginary.}$$

The sign of (ib_j) decides whether the eigenvalue $\varkappa_j$ corresponds to a soliton or an antisoliton. The eigenvalue $\varkappa_j$ itself specifies the soliton or antisoliton momentum.

Multi-breather solutions also correspond to reflectionless potentials. Here the zeros of $a(\varkappa,\eta) = a(\varkappa,o)$ lie symmetrically about, but off the imaginary axis.

There is a 1:1 correspondence between multi-soliton, multi-antisoliton and multi-breather solutions on the one hand and reflectionless potentials on the other hand.

II. 3. The Inverse Scattering Method and the Cauchy Initial Value Problem

Actually, we are not so much interested in the characteristic initial value problem as in the Cauchy initial value problem. To "solve" the latter problem by means of the inverse scattering method, we form appropriate linear combinations of the ξ and η-evolution equations

$$L\psi = \varkappa\psi$$

$$\frac{\partial}{\partial\eta}\psi = - M\psi \quad .$$

Subsequently we change the basis

$$\psi = \frac{1}{\sqrt{2}}\begin{pmatrix}1 & i\\ i & 1\end{pmatrix}\Psi \quad .$$

Thus we arrive at the following isospectral one-parameter family of linear eigenvalue problems with t as the deformation parameter

$$- i\sigma^2\frac{d}{dx}\Psi + \{\frac{i}{4}\, w\sigma^1 + \frac{B^2}{\lambda}\}\Psi = \lambda\Psi$$

$$\frac{\partial}{\partial t}\Psi = \frac{d}{dx}\Psi + 2i\sigma^2\,\frac{B^2}{\lambda}\,\Psi$$

where

$$w = \breve{u}_t + \breve{u}_x$$

$$B = \exp\,(\frac{i}{2}\,\breve{u}\sigma^3)$$

$\sigma^1,\sigma^2,\sigma^3(,\sigma^4)$ are the 3 Pauli (2x2 unit) matrices

and

$$\lambda = \frac{i}{4\zeta} = \frac{\varkappa}{2} \quad .$$

These equations contain the eigenvalue λ also in the denominator. They should be viewed as descendants of the isospectral family of degenerate linear eigenvalue problems (abuse of notation!)

$$L(t)\binom{\Psi}{\chi} = \lambda\binom{\Psi}{\chi}$$

$$\frac{\partial L}{\partial t} = [L,M]$$

$$L = \begin{pmatrix} -i\sigma^2, o \\ o \ , o \end{pmatrix} \frac{d}{dx} + \begin{pmatrix} \frac{i}{4}w\sigma^1, B \\ B \ \ , o \end{pmatrix}$$

$$M = \begin{pmatrix} -\sigma^4, o \\ o \ , \sigma^4 \end{pmatrix} \frac{d}{dx} + \begin{pmatrix} o, -2i\sigma^2 B \\ -2iB\sigma^2, o \end{pmatrix} .$$

These problems are not hermitean with respect to the scalar product

$$\int_{-\infty}^{+\infty} dx\{\Psi^{(1)\dagger}(x)\Psi^{(2)}(x) + \chi^{(1)\dagger}(x)\chi^{(2)}(x)\}$$

where the symbol $\dagger$ denotes hermitean conjugation. However, they are hermitean with respect to the scalar product

$$\int_{-\infty}^{\infty} dx\{\Psi^{(1)T}(x)\Psi^{(2)}(x) + \chi^{(1)T}(x)\chi^{(2)}(x)\}$$

where the symbol T denotes transposition.
We shall consider the eigenvalue problems in the space of functions with a finite L_2-norm based on the first scalar product

$$\int_{-\infty}^{+\infty} dx\{\Psi^{\dagger}(x)\Psi(x) + \chi^{\dagger}(x)\chi(x)\} .$$

Still, in the orthonormality and completeness relations below we shall recognize the second scalar product which for bound state or continuum solutions reads

$$\int_{-\infty}^{+\infty} dx\Psi^T(x,\lambda) \left[\sigma^4 + \frac{B^2(x)}{\lambda\mu}\right] \Psi(x,\mu).$$

Here as in many places below we suppress the t-dependence.

We assume that the Cauchy data u(x) and $u_t(x)$ satisfy the following condition

$$\int_{-\infty}^{+\infty} dx\{|\breve{u}_t(x)| + |\breve{u}_x(x)| + |\sin(\frac{\breve{u}(x)}{2})|\} < \infty$$

which reproduces itself in the course of time.

The chain of linear problems can be depicted schematically as follows:

$o:\{\check{u}(o,x),\check{u}_t(o,x)\}$ ⟶ $L(o)\psi(\circ;\lambda,o)=\lambda\psi(\circ;\lambda,o)$

direct scattering problem

$s(o)=\{r(\lambda,o)\ -\infty<\lambda<+\infty;\ \zeta_j,m_j(o)\ j=1,\dots,N\}$

time evolution of the scattering data

$s(t)=\{r(\lambda,t)\ -\infty<\lambda<+\infty;\ \zeta_j,m_j(t)\ j=1,\dots,N\}$

inverse scattering problem

$t:\{\check{u}(t,x),\check{u}_t(t,x)\}$ ⟵ $L(t)\psi(\circ;\lambda,t)=\lambda\psi(\circ;\lambda,t)$

The fundamental solution matrix of the free eigenvalue problem

$$w = o,\ B = \tfrac{1}{4}\sigma^4$$

is

$$E(x,\lambda)=(e_1(x,\lambda),e_2(x,\lambda))=(e^{i\nu x}\binom{1}{i},e^{-i\nu x}\binom{1}{-i})$$

where

$$\nu = \lambda - \frac{1}{16\lambda}\ ,\ \mathrm{Im}\nu = \mathrm{Im}\lambda\cdot\left(1+\frac{1}{16|\lambda|^2}\right).$$

We see that both e_1 and e_2 as functions of λ are analytic in the entire plane with the exception of the point $\lambda=0$ where they have an essential singularity. For $x\geq 0$ $e_1(x,\lambda)$ is still continuous in the closed upper-half λ-plane, $e_2(x,\lambda)$ in the closed lower-half λ-plane.

The solution matrices

$$F(x,\lambda)=(f_1(x,\lambda),f_2(x,\lambda)),G(x,\lambda)=(g_1(x,\lambda)g_2(x,\lambda))$$

- the solutions f_i are normalized at the right end of the real line, the solutions g_j at the left end - are connected by the transition matrix $T(\lambda)$

$$F(x,\lambda)=G(x,\lambda)T(\lambda), \quad \det T(\lambda)=1 .$$

As a consequence of the symmetries of the problem, $T(\lambda)$ is unitary

$$T(\lambda) = \begin{pmatrix} a(\lambda), & -b^*(\lambda) \\ b(\lambda), & a^*(\lambda) \end{pmatrix}$$

for real λ:

$$|a(\lambda)|^2 + |b(\lambda)|^2 = 1$$

and

$$a(-\lambda)^* = a(\lambda), \ -b(-\lambda)^* = b(\lambda).$$

With the help of orthonormality relations of the type

$$\delta(\mu-\lambda) = \frac{1}{4\pi}\int_{-\infty}^{+\infty} dy\ e_2^T(y,\mu)[\sigma^4 + \frac{\sigma^4}{16\lambda\mu}]e_1(y,\lambda)$$

we derive the following integral representations for F and G

$$F(x,\lambda)=E(x,\lambda)+\int_x^\infty dx\,\mathfrak{A}_1(x,y)E(y,\lambda)+\frac{1}{\lambda}\int_x^\infty dy\,\mathfrak{A}_2(x,y)E(y,\lambda)$$

$$G(x,\lambda)=E(x,\lambda)+\int_{-\infty}^x dy\,\mathfrak{B}_1(x,y)E(y,\lambda)+\frac{1}{\lambda}\int_{-\infty}^x dy\,\mathfrak{B}_2(x,y)E(y,\lambda).$$

We have converted analyticity and growth properties into support properties (Paley-Wiener theorem).

The above representations correspond to the expression for $g_2(x,k)$ previously discussed in the context of the direct scattering problem for a self-adjoint second order differential operator.

The pairs $\mathfrak{A}_1(x,y)$, $\mathfrak{A}_2(x,y)$ and $\mathfrak{B}_1(x,y)$, $\mathfrak{B}_2(x,y)$ satisfy certain linear integro-differential equations, the kernels of which are determined by the Cauchy data. These equations are obtained by inserting the above representations into the integral equations which are equivalent to the differential equation combined with the boundary conditions for F and G, respectively.

By forming certain linear combinations of the entries of $\mathcal{L}_1(x,x)$ and $\mathcal{L}_2(x,x)$, the "potentials" B(x) and $\frac{i}{4}w(x)\sigma^1$ are reconstructed and thereby the Cauchy data recovered.

In order to find the initial scattering data

$$s(o) = \{r(\lambda,o) \quad -\infty<\lambda<+\infty; \zeta_j, \ m_j(o) \quad j=1,\dots,N\}$$

we compute $\mathcal{L}_1(x,y)$ and $\mathcal{L}_2(x,y)$ from the integro-differential equations corresponding to the Cauchy initial data. We then insert the result of the computation into the expressions for $a(\lambda)$ and $b(\lambda)$ in terms of $\mathcal{L}_1$ and $\mathcal{L}_2$.

The scattering data evolve in time like

$$s(t)=\{\exp[-2i(\lambda+\frac{1}{16\lambda})t] \cdot r(\lambda,o) -\infty<\lambda<+\infty; \zeta_j,$$

$$m_j(t)=\exp[-2i(\zeta_j+\frac{1}{16\zeta_j})t]\cdot m_j(o) \quad j=1,..,N\} .$$

It is left as an exercise to the audience to prove these equations.

In order to pass from the scattering data s(t) back to the Cauchy data at time t we have to solve the inverse scattering problem for the linear differential operator under consideration.

It follows from the equation

$$a(\lambda) = \frac{i}{2}\det(f_1(x,\lambda), g_2(x,\lambda))$$

- true for any real value of x - and the fact that $e_1(x,\lambda)$ and $e_2(x,\lambda)$ control the behaviour of $f_1(x,\lambda)$ and $g_2(x,\lambda)$ as λ approaches zero from the upper-half λ-plane (consult the integral representation above) that $a(\lambda)$ can be continued to a function which is analytic (continuous) in the (closed) upper half plane.

$$a(\lambda) \underset{\substack{|\lambda|\to\infty \\ \lambda\geq o}}{=} 1 + o(1)$$

Actually, with $a(\lambda)$ also $a(-\lambda^*)^*$ is an analytic function in the upper half λ-plane. It coincides with $a(\lambda)$ on the real axis. Hence for $\mathrm{Im}\lambda\geq 0$

$$a(\lambda) = a(-\lambda^*)^* .$$

From this we infer that $a(\lambda)$ is real on the imaginary axis and that the zeros of $a(\lambda): =\zeta_j$ lie symmetrically with respect to the imaginary axis.

We shall assume that $a(\lambda)$ does not have zeros on the real axis. This assumption together with the previous bound on the Cauchy data guarantees that there is only a finite number of zeros of $a(\lambda)$ in the upper-half λ-plane, all of which, moreover, are of finite order. (According to ref.[8] the zeros can only accumulate to points of the continuous spectrum).

In addition, for our convenience, we shall assume that all the zeros of $a(\lambda)$ in the upper half plane $\lambda=\zeta_j$ $j=1,..,N$ are simple:

$$a(\lambda)= \dot{a}(\zeta_j)(\lambda-\zeta_j) + O(|\lambda-\zeta_j|^2), \quad \dot{a}(\zeta_j) \neq 0.$$

For $\lambda=\zeta_j$, the solutions f_1 and g_2 become linearly dependent

$$f_1(x,\zeta_j) = b_j g_2(x,\zeta_j).$$

The solution f_1 decreases exponentially at both ends of the real x-axis. Thus, the zeros ζ_j $j=1,..,N$ are the discrete eigenvalues of the operator L.

We set

$$m_j = \frac{b_j}{i\dot{a}(\zeta_j)} \quad .$$

We enumerate first the zeros on the imaginary axis

$$\zeta_\ell = i\varkappa_\ell \qquad \ell = 1,..,n_1$$

and then the zeros off the imaginary axis which come in pairs

$$\zeta_{n_1+2k-1} = \lambda_k, \ \zeta_{n_1+2k} = -\lambda_k^* \qquad k = 1,..,n_2$$

$$0 < \arg \lambda_k < \frac{\pi}{2}$$

$$n_1 + 2n_2 = N \quad .$$

We solve the inverse scattering problem in a similar fashion as used for the self-adjoint second-order differential operator. We write down the equation defining the coefficients $a(\lambda)$ and $b(\lambda)$ in the following modified form

$$\left[\frac{1}{a(\lambda)} - 1\right] f_1(x,\lambda) = r(\lambda)[g_2(x,\lambda) - e_2(x,\lambda)] + r(\lambda) e_2(x,\lambda) + [g_1(x,\lambda) - e_1(x,\lambda)] - [f_1(x,\lambda) - e_1(x,\lambda)].$$

Correspondingly,

$$\left[\frac{1}{a(\lambda)^*} - 1\right] f_2(x,\lambda) = \ldots$$

We multiply the first equation by $e_2^T(y,\lambda)/4\pi\,(e_2^T(y,\lambda)/4\pi\lambda)$ and the second equation by $e_1^T(y,\lambda)/4\pi \quad (e_1^T(y,\lambda)/4\pi\lambda)$ and integrate the sum for $x > y$ over the entire real λ-axis. We evaluate the l.h.s. by the residue theorem in the upper and lower half planes, respectively, and insert the integral representation for $G(x,\lambda)$. These manipulations yield the Gelfand-Levitan-Marchenko equations

$$x>y:\quad 0 = K_1(x,y) + \mathcal{F}_1(x,y) + \int_{-\infty}^{x} du\, K_1(x,u)\mathcal{F}_1(u,y) + \int_{-\infty}^{x} du\, K_2(x,u)\,\mathcal{F}_2(u,y)$$

$$0 = 16K_2(x,y) + \mathcal{F}_2(x,y) + \int_{-\infty}^{x} du\, K_1(x,u)\mathcal{F}_2(u,y) + \int_{-\infty}^{x} du\, K_2(x,u)\,\mathcal{F}_3(u,y)$$

whose kernels $\mathcal{F}_\ell(x,y)\quad \ell=1,2,3$ are determined by the scattering data's through

$$\mathcal{F}_\ell(x,y) = \frac{1}{4\pi}\int_{\infty}^{\infty} \frac{d\lambda}{(\lambda)^{\ell-1}}\{r(\lambda)e_2(x,\lambda)e_2^T(y,\lambda) - r(\lambda)^* e_1(x,\lambda)e_1^T(y,\lambda)\} + \frac{1}{2}\sum_{j=1}^{N}\left\{\frac{m_j}{(\zeta_j)^{\ell-1}}\, e_2(x,\zeta_j)e_2^T(y,\zeta_j) - \frac{m_j^*}{(\zeta_j^*)^{\ell-1}}\, e_1(x,\zeta_j^*)e_1^T(x,\zeta_j^*)\right\}$$

and whose solutions $K_1(x,y)$ and $K_2(x,y)$ are essentially identical with $\mathcal{L}_1(x,y)$ and $\mathcal{L}_2(x,y)$. The relation of $K_1(x,x)$, $K_2(x,x)$ on one hand and the "potentials" on the other hand is given by

$$[K_1(x,x),\ \sigma^2] = -\frac{1}{4}\,w(x)\sigma^1$$

$$-i\sigma^2 K_2(x,x)+iB^2(x)K_2(x,x)16\sigma^2+B^2(x)-\frac{\sigma^4}{16} = 0\ .$$

Thus, in order to pass from the scattering data s(t) back to the Cauchy data at time t - and thereby perform the last step to solve the Cauchy initial value problem - we compute $K_1(x,y)$ and $K_2(x,y)$ from the above linear integral equations for s=s(t) and determine u(t,x) and $u_t(t,x)$ from $K_1(x,x)$ and $K_2(x,x)$ according to the last two relations.

II. 4. The Sine Gordon Theory in Action Angle Variables

In this section we shall study the sine Gordon theory by means of the Hamiltonian formalism in order to determine the elementary excitations of the system [9].

We take u and $\frac{1}{\gamma}u_t$ as generalized coordinates and momenta to start with. The Hamiltonian, i.e. the energy of the system, is

$$P_0 = \frac{1}{2\gamma}\int_{-\infty}^{+\infty} dx\ \{u_t^2 + u_x^2 + 2(1-\cos u)\}\ ,$$

its momentum

$$P_1 = -\frac{1}{\gamma}\int_{-\infty}^{+\infty} dx\ u_x \cdot u_t$$

and its symplectic form

$$\Omega = \frac{1}{\gamma}\int_{\infty}^{\infty} dx\ \{du_t(x) \wedge du(x)\}\ .$$

The invariance of Ω is the criterion for a transformation to be canonical.

The passage from the generalized coordinates u and $\frac{1}{\gamma}u_t$ to the scattering data of the degenerate linear differential operators L(t) provides a canonical transformation to action angle variables. In these variables the Hamiltonian and momentum are cyclic with respect to the angles, and the equations of motion are easily integrated (compare the above exercise).

To show this we first aim to express Ω in terms of the scattering data. We determine certain combinations of the scattering data which leave the external differential form Ω invariant and introduce them as new generalized coordinates. Then without proof we write down P_0 and P_1 in terms of the new coordinates. The expressions for P_0 and P_1 suggest an interpretation of the field theory in terms of particles.

We start by comparing two scattering problems characterized by the linear differential operators $L^{(1)}$ and $L^{(2)}$ or equivalently by the corresponding scattering data $s^{(1)}$ and $s^{(2)}$. Actually, in the end we shall only be interested in comparing scattering problems which are infinitesimally "close" to each other. By a similar reasoning as before we derive integral representations for $F^{(2)}(x,\lambda)$ and $G^{(2)}(x,\lambda)$, the fundamental solution matrices of problem (2), in terms of $F^{(1)}(x,\lambda)$ and $G^{(1)}(x,\lambda)$, the fundamental solution matrices of problem (1):

$$F^{(2)}(x,\lambda) = F^{(1)}(x,\lambda) + \int_x^\infty dy\, \hat{\mathcal{A}}_1(x,y)F^{(1)}(y,\lambda) + \frac{1}{\lambda}\int_x^\infty dy\, \hat{\mathcal{A}}_2(x,y)F^{(1)}(y,\lambda)$$

$$G^{(2)}(x,\lambda) = G^{(1)}(x,\lambda) + \int_\infty^x dy\, \mathcal{L}_1(x,y)G^{(1)}(y,\lambda) + \frac{1}{\lambda}\int_\infty^x dy\, \mathcal{L}_2(x,y)G^{(1)}(y,\lambda) \; .$$

These integral representations are inserted into the equations

$$\frac{1}{4\pi\lambda^\ell}\{[\frac{1}{a^{(2)}(\lambda)} - 1]\, f_1^{(2)}(x,\lambda) - [\frac{1}{a^{(1)}(\lambda)} - 1]\, f_1^{(1)}(x,\lambda)\}\, g_2^{(1)T}(y,\lambda)$$
$$+ \frac{1}{4\pi\lambda^\ell}\{[\frac{1}{a^{(2)}(\lambda)^*} - 1]\, f_2^{(2)}(x,\lambda) - [\frac{1}{a^{(1)}(\lambda)^*} - 1]\, f_2^{(1)}(x,\lambda)\}\, g_1^{(1)T}(y,\lambda) = \ldots$$

$\ell = 0,1$. Subsequently, for $x > y$ both sides of the equations are integrated over the entire real λ-axis. When the integrals are evaluated we obtain the following linear equations (GLM)

$$x>y:\ 0=K_1(x,y)+\mathcal{F}_1(x,y)+\int_{-\infty}^{x}du\ K_1(x,u)\mathcal{F}_1(u,y)+\int_{-\infty}^{x}du\ K_2(x,u)\mathcal{F}_2(u,y)$$

$$0=K_2(x,y)[B^{(1)}(y)]^{-2}+\mathcal{F}_2(x,y)+\int_{-\infty}^{x}du\ K_1(x,u)\mathcal{F}_2(u,y)+\int_{-\infty}^{x}du\ K_2(x,u)\mathcal{F}_3(u,y)$$

the kernels of which are determined by the scattering data $s^{(1)}$ and $s^{(2)}$ through

$$\mathcal{F}_\ell(x,y)=\frac{1}{4\pi}\int_{-\infty}^{+\infty}d\lambda\ \left\{\frac{r^{(2)}(\lambda)-r^{(1)}(\lambda)}{\lambda^{\ell-1}}\ g_2^{(1)}(x,\lambda)\ g_2^{(1)T}(y,\lambda) - \frac{r^{(2)}(\lambda)^*-r^{(1)}(\lambda)^*}{\lambda^{\ell-1}}\ g_1^{(1)}(x,\lambda)\ g_1^{(1)T}(y,\lambda)\right\}$$

$$+\frac{1}{2}\sum_{j=1}^{N^{(2)}}\left\{\frac{m_j^{(2)}}{(\zeta_j^{(2)})^{\ell-1}}\ g_2^{(1)}(x,\zeta_j^{(2)})\ g_2^{(1)T}(y,\zeta_j^{(2)}) - \frac{m_j^{(2)*}}{(\zeta_j^{(2)*})^{\ell-1}}\ g_1^{(1)}(x,\zeta_j^{(2)*})\ g_1^{(1)T}(y,\zeta_j^{(2)*})\right\}$$

$$-\frac{1}{2}\sum_{j=1}^{N^{(1)}}\{m_j^{(2)}\to m_j^{(1)}\ ,\ \zeta_j^{(2)}\to\zeta_j^{(1)}\}\qquad \ell = 1,2,3.$$

$K_1(x,y)$ and $K_2(x,y)$ are essentially identical with $\mathcal{L}_1(x,y)$ and $\mathcal{L}_2(x,y)$ whence

$$[K_1(x,x),\ \sigma^2] = -\ \frac{1}{4}(w^{(2)}(x)-w^{(1)}(x))\sigma^1$$

$$-i\sigma^2K_2(x,x)+i[B^{(2)}(x)]^2K_2(x,x)[B^{(1)}(x)]^{-2}\sigma^2+[B^{(2)}(x)]^2 -[B^{(1)}(x)]^2 = 0.$$

Note that $\mathcal{F}_\ell(x,y)$ ℓ =1,2,3 and consequently $K_i(x,y)$ i=1,2 are of first order in infinitesimal changes of the scattering data

$$r^{(1)}(\lambda) = r(\lambda) \;;\; \zeta_j^{(1)} = \zeta_j,\ m_j^{(1)} = m_j \quad j=1,..,N^{(1)}=N$$

$$r^{(2)}(\lambda)=r(\lambda)+dr(\lambda);\zeta_j^{(2)}=\zeta_j+d\zeta_j,m_j^{(2)}=m_j+dm_j \quad j=1,..,N^{(2)}=N.$$

Thus, keeping only first order terms, the GLM equations are solved by

$$K_1(x,y) = -\mathcal{F}_1(x,y) \ ,\ K_2(x,y) = -\mathcal{F}_2(x,y)$$

and we obtain

$$-\frac{\sigma^1}{4}\, dw(x) = [\sigma^2, \mathcal{F}_1(x,x)]$$

$$-\sigma^3 du(x)= \sigma^2\, \mathcal{F}_2(x,x)[B(x)]^{-2} - [16B(x)]^2\, \mathcal{F}_2(x,x)\sigma^2$$

or

$$du(x)=\frac{i}{\pi} \int_{-\infty}^{+\infty} d\lambda dr(\lambda)g(x,\lambda)+2i\sum_1^N\{g(x,\zeta_j)dm_j + m_j(\frac{\partial}{\partial\lambda}g(x,\lambda)/_{\lambda=\zeta_j}d\zeta_j\}$$

$$dw(x)=\frac{2i}{\pi} \int_{\infty}^{\infty} d\lambda dr(\lambda)f(x,\lambda)+4i\sum_1^N\{f(x,\zeta_j)dm_j + m_j(\frac{\partial}{\partial\lambda}f(x,\lambda)/_{\lambda=\zeta_j}d\zeta_j\} \quad .$$

Here $g(x,\lambda)$ and $f(x,\lambda)$ are certain quadratic forms in the components of the fundamental solution $g_2^{(1)}(x,\lambda)$.

We insert this result into the expression for Ω and evaluate the remaining x-integration with the help of the differential equation satisfied by $g_2^{(1)}(x,\lambda)$. The integrand turns out to be a complete derivative with respect to x. Thus, the integral can be immediately expressed in terms of the asymptotic behaviour of $g_2^{(1)}(x,\lambda)$, i.e. in terms of the scattering data. Yet, in the differentials $dr(\lambda)$, $d\zeta_j$ and dm_j Ω does not have the symplectic normal form. Moreover, since the λ-integration extends from $-\infty$ to $+\infty$, both differentials $dr(-\lambda)$ and $dr(\lambda)$ occur which are not independent.

However, the combinations

$$\rho(\lambda)= -\frac{8}{\pi\gamma\lambda}\, \ell n|a(\lambda)|, \quad \phi(\lambda) = -\arg b(\lambda) \qquad \lambda > 0$$

$$p_\ell=\frac{1}{\gamma}\ \ell n\varkappa_\ell,\quad q_\ell= 8\ \ell n|c_\ell| \quad \ell=1,..,n_1 \quad c_\ell=-ib_\ell=+m_\ell\dot{a}(i\varkappa_\ell)$$

$$\xi_k=\frac{4}{\gamma}\ell n|\lambda_k|\ ,\quad \eta_k= 4\ell n|d_k|$$

$$k=1,..,n_2 \quad d_k=m_k\cdot\dot{a}(\lambda_k)$$

$$\Theta_k=\arg\lambda_k,\quad \phi_k=-\frac{16}{\gamma}\arg d_k$$

are canonical variables in which Ω adopts the form

$$\Omega=\int_0^\infty d\lambda d\rho(\lambda)\wedge d\phi(\lambda)+\sum_{\ell=1}^{n_1}dp_\ell\wedge dq_\ell+\sum_{k=1}^{n_2}(d\xi_k\wedge d\eta_k+d\Theta_k\wedge d\phi_k)\ .$$

We express the energy and the momentum of the system in terms of the new variables with the help of the so-called trace identities [9,1o]:

$$P_{\substack{o\\1}} = P^{(1)}_{\substack{o\\1}} + P^{(2)}_{\substack{o\\1}} + P^{(3)}_{\substack{o\\1}}$$

where

$$P^{(1)}_{\substack{o\\1}} = \int_0^\infty d\lambda\rho(\lambda)\ (\frac{1}{8\lambda}\pm 2\lambda)$$

$$P^{(2)}_{\substack{o\\1}} = \frac{1}{\gamma}\sum_{\ell=1}^{n_1}(\frac{1}{\varkappa_\ell}\pm 16\varkappa_\ell)$$

$$P^{(3)}_{\substack{o\\1}} = \frac{2}{\gamma}\sum_{k=1}^{n_2}\ \sin\Theta_k(\frac{1}{|\lambda_k|}\pm 16|\lambda_k|).$$

P_o and P_1 are cyclic with respect to the angle variables $\phi(\lambda)$, q_ℓ, η_k and ϕ_k. Hence, the actions $\rho(\lambda)$, p_ℓ,ξ_k and Θ_k are constant in time and the angles vary only linearly with time.

The form of P_o and P_1 suggests a simple interpretation of the sine Gordon field in terms of particles:

1) For fixed λ, the variable $\rho(\lambda)$ takes only positive values, the variable $\phi(\lambda)$ values on the interval $[0,2\pi]$, i.e. on a circle. Thus, for fixed $\lambda,\rho(\lambda)$ and $\phi(\lambda)$ form a pair of canonical variables of the type of a particle number and phase. Hence, we interpret

$$\rho(\lambda) \qquad \text{as the density,}$$

$$p(\lambda) = \frac{1}{8\lambda} - 2\lambda \qquad \text{as the momentum,}$$

$$h(\lambda) = \frac{1}{8\lambda} + 2\lambda \qquad \text{as the energy, and}$$

$$\sqrt{h^2(\lambda)-p^2(\lambda)} = 1 \qquad \text{as the mass}$$

of particles of type 1.

2) $P_o^{(2)}$ and $P_1^{(2)}$ are already written in the form of a sum over particles with

$$\text{momentum} \quad \frac{1}{\gamma}\left(\frac{1}{\varkappa_\ell} - 16\,\varkappa_\ell\right),$$

$$\text{energy} \quad \frac{1}{\gamma}\left(\frac{1}{\varkappa_\ell} + 16\,\varkappa_\ell\right) \text{ and}$$

$$\text{mass} \quad \sqrt{\left[\frac{1}{\gamma}\left(\frac{1}{\varkappa_\ell} + 16\,\varkappa_\ell\right)^2 - \frac{1}{\gamma}\left(\frac{1}{\varkappa_\ell} - 16\,\varkappa_\ell\right)\right]^2} = \frac{8}{\gamma}.$$

Thus, we identify the ℓth particle of type 2 with a soliton if $c_\ell < 0$, and with an antisoliton if $c_\ell > 0$.

3) $P_o^{(3)}$ and $P_1^{(3)}$ are sums of contributions from particles with an internal degree of freedom; the corresponding phase spaces are four-dimensional. The particles have

$$\text{momentum} \quad \frac{2\sin\Theta_k}{\gamma}\left(\frac{1}{|\lambda_k|} - 16|\lambda_k|\right),$$

$$\text{energy} \quad \frac{2\sin\Theta_k}{\gamma}\left(\frac{1}{|\lambda_k|} + 16|\lambda_k|\right) \text{ and}$$

$$\text{mass} \quad M_k = \frac{16}{\gamma}\sin\Theta_k < 2\cdot\frac{8}{\gamma}$$

We identify the particles of type 3 with the breathers.

Thereby, we have determined the elementary excitations of the classical sine Gordon theory. With their help, the possible trajectories of the system are conveniently described.

III. The Mass Spectrum of the Fully Quantized sine Gordon Theory

In the course of this third and last lecture a rather

sketchy review of the recent work by A.Luther shall be given [11] . Luther succeeded in calculating the energy momentum spectrum of the fermion antifermion bound states in the quantized renormalized Thirring model. In view of the equivalence of the quantized massive Thirring model and the sine Gordon theory for a certain range of the respective coupling constants [12,13] this yields exact expressions for the masses of the sine Gordon breathers.

The following exposition owes a great deal to M. Lüscher who gave a very clear discussion of the main ideas involved in Luther's approach and developing these ideas further derived very restrictive properties of the soliton, antisoliton, breather reactions [14] .

Luther obtained information about the fully quantized sine Gordon theory along the following line:
a) eight-vertex model $\longrightarrow$ b) the XYZ model (anisotropic Heisenberg spin chain) $\longrightarrow$ c) lattice variant of the massive Thirring model $\longrightarrow$ d) massive Thirring model $\longrightarrow$ e) sine Gordon theory.

We shall now start by describing the models involved.

III.1. The Models

a) The eight-vertex model

Consider a lattice of M rows and N columns, M and N being even integers, i.e., N=2r. Draw arrows between the lattice sites such that an even number of arrows points into each of them. Impose cyclic (+) or anticyclic (-) boundary conditions on the horizontal arrows. At every lattice site exactly eight configurations of arrows are possible.

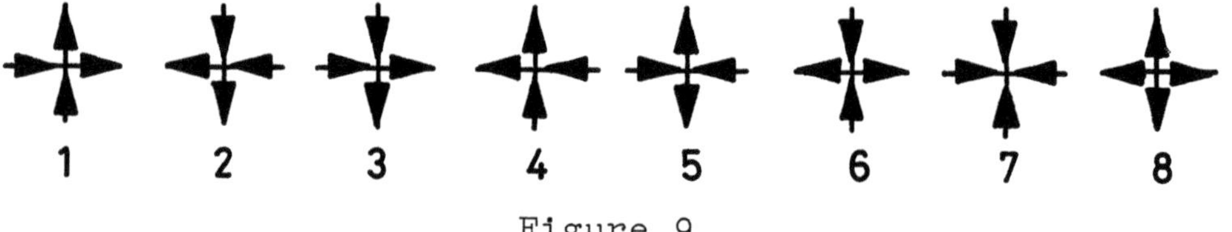

Figure 9

Associate the energy ε_j

$$\varepsilon_1 = \varepsilon_2, \ \varepsilon_3 = \varepsilon_4, \ \varepsilon_5 = \varepsilon_6, \ \varepsilon_7 = \varepsilon_8$$

to each "vertex" of type j $j = 1,..,8$.

The partition function $Z^{(\pm)}$ is defined by

$$Z^{(\pm)} = \sum \exp \{-\beta \sum_{1}^{8} N_j \varepsilon_j\}$$

where the summation extends over all allowed configurations of arrows on the lattice and where N_j is the total number of vertices of type j.

Consider $Z^{(\pm)}$ as a function of the following four linear combinations of the Boltzmann weights

$$w_1 = \frac{1}{2}(e^{-\beta\varepsilon_6} + e^{-\beta\varepsilon_8}), \qquad w_2 = \frac{1}{2}(e^{-\beta\varepsilon_6} - e^{-\beta\varepsilon_8})$$

$$w_3 = \frac{1}{2}(e^{-\beta\varepsilon_2} - e^{-\beta\varepsilon_4}), \qquad w_4 = \frac{1}{2}(e^{-\beta\varepsilon_2} + e^{-\beta\varepsilon_4}) .$$

The partition function can be expressed as a trace involving the Mth power of the transfer matrix

$$\mathbb{T}^{(\pm)} = \mathbb{T}^{(\pm)}(w_1,\ldots,w_4) .$$

The entries of the transfer matrix for any two successive rows of vertical arrows are indexed by the corresponding arrow configurations $\underline{\alpha}$ and $\underline{\alpha'}$

$$\underline{\alpha}^{(\prime)} = \{\alpha_{-r+1}^{(\prime)},\ldots,\alpha_r^{(\prime)}\}, \alpha_j^{(\prime)} = \begin{cases} +, & \text{the arrow in column } j \text{ points up} \\ -, & \text{the arrow in column } j \text{ points down} \end{cases}$$

The entries of the 2^N by 2^N matrix are

$$\mathbb{T}^{(\pm)}_{\underline{\alpha},\underline{\alpha}} = \sum \exp(-\beta \sum_{1}^{8} n_j \varepsilon_j)$$

where the sum extends over all allowed arrangements of arrows in the intervening row of horizontal arrows and where n_j is the number of vertices of type j in this row.

Actually, the entries of the transfer matrix $\mathbb{T}^{(\pm)}$ in their turn can be represented as traces of N-fold products of two by two matrices $\mathbb{R}(\alpha_j,\alpha_j')$

$$\mathbb{T}^{(\pm)}_{\underline{\alpha},\underline{\alpha}'} = \{\mathbb{R}(\alpha_{-r+1}, \alpha'_{-r+1}) \ldots \mathbb{R}(\alpha_r, \alpha'_r)\}$$

$$\mathbb{T}^{(-)}_{\underline{\alpha},\underline{\alpha}} = \{\mathbb{R}(\alpha_{-r+1}, \alpha'_{-r+1}) \ldots \mathbb{R}(\alpha_r, \alpha'_r) i\sigma^1\}$$

with

$$\mathbb{R}\,(\alpha_j, \alpha'_j)_{\lambda\lambda'} = \sum_{\varkappa=1}^{4} w_\varkappa \, \sigma^\varkappa_{\alpha_j\alpha'_j} \sigma^\varkappa_{\lambda\lambda'}$$

As before, the symbols $\sigma^\varkappa$ $\varkappa=1,..,4$ stand for the three Pauli matrices and the two by two unit matrix respectively.

The matrix $\mathbb{T}^{(\pm)}$ commutes with the operator

$$\mathcal{C}'' = \prod_{-r+1}^{r} \sigma^1_j$$

$\mathcal{C}''$ is equal to $(-1)^{\nu''}$ when acting on a state vector which is symmetric (ν'' even) or antisymmetric (ν'' odd) with respect to the reversal of all arrows. Without loss of generality we set w_3 equal to one. The remaining weights w_1, w_2 and w_4 are expressed in terms of three new parameters V, ℓ, ζ

$$w_1 = \frac{\mathrm{cn}(V,\ell)}{\mathrm{cn}(\zeta,\ell)}, \quad w_2 = \frac{\mathrm{dn}(V,\ell)}{\mathrm{dn}(\zeta,\ell)}, \quad w_3 = 1, \quad w_4 = \frac{\mathrm{sn}(V,\ell)}{\mathrm{sn}(\zeta,\ell)}$$

where $\mathrm{sn}(u,\ell)$, $\mathrm{cn}(u,\ell)$ and $\mathrm{dn}(u,\ell)$ denote the Jacobian elliptic functions of argument u and modulus ℓ. We restrict our attention to the so-called fundamental region

$$0 < \ell < 1, \quad 0 < \zeta < K_\ell$$

where K_ℓ stands for the complete elliptic integral of the first kind of modulus ℓ. After having fixed the values of ℓ and ζ in this domain, we consider $\mathbb{T}^{(\pm)}(w_1 \ldots, w_4)$ as a function of V: $\mathbb{T}^{(\pm)}(V)$.

Baxter has shown that the matrices $\mathbb{T}^{(\pm)}(V)$ are normal, consequently diagonalizable, and that $\mathbb{T}^{(\pm)}(V)$ and $\mathbb{T}^{(\pm)}(V')^{(+)}$ commute for arbitrary values of V and V'. Hence the matrices $\mathbb{T}^{(\pm)}(V)$ can be diagonalized for all Vs simultaneously, the eigenvectors are V-independent, the eigenvalues entire functions of V [15] .

In the limit $N\to\infty$ the largest eigenvalue of the transfer matrix $\mathbb{T}^{(+)}(V)$: $T_0^{(+)}(V)$ was calculated by Baxter [15] , in the same limit the next largest eigenvalues of $\mathbb{T}^{(+)}(V)$:$T_B^{(+)}(V)$ and the largest eigenvalues of $\mathbb{T}^{(-)}(V)$: $T_S^{(-)}(V)$ and $T_A^{(-)}(V)$ were computed by Johnson, Krinsky and McCoy [16] , the latter ones only implicitly. It is this information which has to be passed down the line to the sine Gordon model.

In order to state the exact results, some more notation has to be introduced:

$$\tau = \frac{\pi K_\ell}{K'_\ell}\ , \quad \lambda = \frac{\pi\zeta}{K'_\ell}\ , \quad \mu = \frac{\pi\zeta}{K_\ell}$$

where

$$K'_\ell = K_{\sqrt{1-\ell^2}}\ .$$

New moduli k_1 and k_2 are defined by the implicit equations

$$\frac{\pi K'_1}{K_1} = \lambda,\quad \frac{\pi K'_2}{K_2} = 2\lambda \qquad K_i = K_{k_i} \quad i = 1,2.$$

Now, the results of Johnson, Krinsky and McCoy read as follows

$$\frac{T_S^{(-)}(V)}{T_0^{(+)}(V)} = \sqrt{k_2}\ \mathrm{sn}[\tfrac{K_2}{\pi}(\phi - i\tfrac{\pi V}{K'_\ell}), k_2] \quad 0 \le \phi \le 2\pi$$

$$\frac{T_A^{(-)}(V)}{T_0^{(+)}(V)} = \sqrt{k_2}\ \mathrm{sn}[\tfrac{K_2}{\pi}(\phi - i\tfrac{\pi V}{K'_\ell}), k_2] \quad -2\pi \le \phi \le 0$$

and for $\mu > \frac{\pi}{2}$

$$\frac{T_B^{(+)}(V)}{T_0^{(+)}(V)} = k_2\ \mathrm{sn}[\tfrac{K_2}{\pi}(\phi_+ - i\tfrac{\pi V}{K'_\ell}),\ k_2]\ \mathrm{sn}\,[\tfrac{K_2}{\pi}(\phi_- - i\tfrac{\pi V}{K'_\ell}),\ k_2]$$

$$\phi_\pm = \phi \pm in(\tau-\lambda) \mp i\lambda\ , \qquad 0 \le \phi \le 2\pi$$

where in addition to the continuous variable ϕ a discrete label $n=1,2,\ldots<[\frac{\pi}{\mu}-1]^{-1}$ occurs.

b) The XYZ model

The Hamiltonian defining the anisotropic Heisenberg spin chain is given by

$$H_{XYZ}^{(\pm)} = -\frac{1}{2}\sum_{-r+1}^{r}\{J_x\sigma_k^x\sigma_{k+1}^x + J_y\sigma_k^y\sigma_{k+1}^y + J_z\sigma_k^z\sigma_{k+1}^z\} .$$

The Pauli matrices σ_k^x, σ_k^y and σ_k^z act on the up and down spin positions at site k, the Hamiltonians $H_{XYZ}^{(\pm)}$ operate on the configurations of up and down spin positions at the 2r sites of the chain.

For $H_{XYZ}^{(+)}$ cyclic boundary conditions are assumed:

$$\sigma_{r+1}^{x(,y,z)} = \sigma_{-r+1}^{x(,y,z)} .$$

For $H_{XYZ}^{(-)}$ anticyclic boundary conditions are assumed:

$$\sigma_{r+1}^{x(,y)} = -\sigma_{-r+1}^{x(,y)} , \quad \sigma_{r+1}^{z} = \sigma_{-r+1}^{z} .$$

$H_{XYZ}^{(+)}$ and $H_{XYZ}^{(-)}$ have different spectra.

We are interested in the regime

$$J_x > J_y > |J_z| .$$

The operator $\prod_{-r+1}^{r}\sigma_k^z$ commutes with $H_{XYZ}^{(+)}$ and $H_{XYZ}^{(-)}$. Its eigenvalues are 1(-1) if the number of down spins F is even (odd).

We introduce the projection operators

$$\Pi_+ = \frac{1}{2}(1-(-1)^{r+F}) , \quad \Pi_- = \frac{1}{2}(1+(-1)^{r+F})$$

and define still another Hamiltonian H by

$$H = \Pi_+ H_{XYZ}^{(+)} + \Pi_- H_{XYZ}^{(-)} + \text{const.}$$

$$H : \left(\begin{array}{c|c} H^{(+)}_{XYZ} + \text{const.} & 0 \\ (-1)^F = (-1)^{r+1} & \\ \hline 0 & H^{(-)}_{XYZ} + \text{const.} \\ & (-1)^F = (-1)^r \end{array} \right)$$

c) The lattice variant of the massive Thirring model

The massive Thirring model on a one-dimensional lattice with spacing a and N=2r lattice sites is defined by the Hamiltonian

$$H_{L.Th.} = \sum_{-r+1} \left\{ \frac{i}{2a} v(G)(\phi^+_n \phi_{n+1} - \phi^+_{n+1} \phi_n) + (-1)^n \frac{m_o}{2}(\phi^+_n \phi^+_{n+1} + \phi_{n+1} \phi_n) - \frac{G}{2a}(\phi^+_n \phi_n - \tfrac{1}{2})(\phi^+_{n+1} \phi_{n+1} - \tfrac{1}{2}) \right\} - E_o .$$

The ϕ_ns are fermion operators

$$\{\phi_n, \phi^+_m\} = \delta_{nm}.$$

Cyclic boundary conditions are assumed: $\phi_{r+1} = \phi_{-r+1}$.

The parameter G plays the rôle of a renormalized coupling constant, the renormalization constant m_o the rôle of a bare mass. The finite factor v(G) must be included in the Hamiltonian in order to obtain the correct dispersion law

$$E = \sqrt{M^2 + p^2} \qquad (c = 1)$$

for the one particle states in the continuum limit. Here the symbol M stands for the renormalized fermion mass. E_o is the ground state energy.

As the lattice spacing a varies, we keep the parameters G and v(G) constant and equal to

$$G = - \frac{4\varepsilon}{\pi} \mathrm{ctg}\, \varepsilon, \; v(G) = \frac{2\varepsilon}{\pi \sin\varepsilon} \qquad 0 < \varepsilon < \pi$$

and let the bare mass m_o depend on a as follows

$$m_o = \frac{8\varepsilon \sin\varepsilon}{\pi a} \left(\frac{aM}{4} \right)^{2\varepsilon/\pi}$$

In the continuum limit $a \downarrow 0$ the leading short distance behaviour of the theory coincides with that one of a massless Thirring model with coupling constant g:

$$g = -\pi \frac{\frac{\pi}{2}-\varepsilon}{\pi-\varepsilon} .$$

d) The massive Thirring model

This is a theory of a single Dirac field $\psi(t,x)$ in one time and one space dimension. The Lagrangian density defining the model is

$$\mathcal{L} = i\bar{\psi}\gamma^{\mu}\partial_{\mu}\psi - M_{o}\psi\psi - \frac{g}{2}(\bar{\psi}\gamma^{\mu}\psi)(\bar{\psi}\gamma_{\mu}\psi)$$

where M_o and g play the rôle of a bare mass and a coupling constant respectively.

e) The (quantum) sine Gordon theory

This is a model of a single scalar field $\phi(t,x)$ in one time and one space dimension. Its dynamics is determined by the Lagrangian density

$$\mathcal{L} = \frac{1}{2}\partial_{\mu}\phi\partial^{\mu}\phi + m_{\circ}^{2}\frac{m^{2}}{g_{s.G.}^{2}}\cos\left(\frac{g_{s.G.}}{m}\phi\right) + \mathcal{L}_{vac.}$$

where m_o is a bare mass, $\frac{g_{s.G.}}{m}$ a coupling constant and $\mathcal{L}_{vac.}$ a parameter needed to fix the ground state energy to zero.

III.2. From the Eight-Vertex Model to the Quantum sine Gordon Theory

In this section we shall discuss the relation of the various models a)-e). In particular, we shall dwell on the connection of the parameters and operators (fields) introduced in the context of the separate models.

(a → b) Relation of the eight-vertex and the XYZ models

The transfer matrices $\mathbb{T}^{(\pm)}(V)$ act on the same state space of the Hamiltonians $H^{(\pm)}_{XYZ}$. B. Sutherland was the first one to notice that $\mathbb{T}^{(\pm)}_{(a,b,c,d)}$ commutes with $H^{(\pm)}_{XYZ}$

provided that J_x, J_y and J_z are related to a,b,c,d in a certain fashion. This observation was followed by Baxter's analysis showing that H^{+}_{XYZ} is proportional to a logarithmic derivative of $\mathbb{T}^{(+)}(V)$:

$$H^{(\pm)}_{XYZ} = J_x \mathrm{sn}(2\zeta,\ell) U[\mathbb{T}^{(\pm)}(V)^{-1} \frac{\partial}{\partial V} \mathbb{T}^{(\pm)}(V)]_{V=\zeta} U^{-1} + \text{const.}$$

if the following identifications are made:

$$\ell^2 = \frac{J_x^2 - J_y^2}{J_x^2 - J_z^2} \,, \quad \mathrm{cn}(2\zeta,\ell) = -{}^{J_z}/{}_{J_x} \ .$$

U denotes a unitary transformation acting separately on the Pauli matrices at each lattice site k:

$$U\sigma_k^x U^{-1} = -\sigma_k^z$$

$$U\sigma_k^y U^{-1} = \begin{cases} -\sigma_k^y & (k \text{ odd}) \\ +\sigma_k^y & (k \text{ even}) \end{cases}$$

$$U\sigma_k^z U^{-1} = \begin{cases} -\sigma_k^x & (k \text{ odd}) \\ +\sigma_k^x & (k \text{ even}) \end{cases} \ .$$

This transformation maps the fundamental region of the eight-vertex model into the J_x, J_y, J_z domain of interest for the XYZ model.

$$U(-1)^{\nu''} U^{-1} = (-1)^F \ .$$

The momentum operator, i.e. (-i)times the logarithm of the operator effecting shifts by one lattice site is given by

$$\mathbb{P}^{(\pm)}_{XYZ} = -iU[\ln(\mathbb{T}^{(\pm)}(\zeta))]U^{-1} + iN\ln 2.$$

(b → c) <u>Relation of the spin chain and the lattice variant of the massive Thirring model</u>

The Jordan Wigner transformation relates the operators ϕ_k and $\sigma_k{}^{x,y,z}$ as follows

$$\phi_k^+ = \exp(i\,\tfrac{\pi}{4}(N+1))\cdot\sigma_k^+ \prod_{-r+1}^{k-1}[i\sigma_j^z], \quad \sigma_k^+ = \tfrac{1}{2}(\sigma_k^x + i\sigma_k^y) .$$

Thereby the Hamiltonians $H_{XYZ}^{(\pm)}$ and $H_{L.Th.}$ are connected according to

$$H_{L.Th.} = H_{XYZ}^{(+)} + \tfrac{1}{2}(1+(-1)^{r+F})\{J_x\sigma_r^x\sigma_{-r+1}^x + J_y\sigma_r^y\sigma_{-r+1}^y\} + \text{const.}$$

$$= \prod_+ H_{XYZ}^{(+)} + \prod_- H_{XYZ}^{(-)} + \text{const.} = H$$

if we identify

$$J_x = \frac{v}{2a} + \frac{m_o}{2}, \quad J_y = \frac{v}{2a} - \frac{m_o}{2}, \quad J_z = \frac{G}{4a} .$$

Clearly

$$J_x > J_y > |J_z|$$

$$(-1)^F = e^{i\pi\sum_{-r+1}^{r}\phi_k^+\phi_k} .$$

The operators $\prod_+$ and $\prod_-$ project on the subspaces of an even and odd number of fermions above the ground state respectively.

The momentum of the S-state and the A-state is

$$p = \frac{q}{a}(\text{mod } \frac{\pi}{a}) \quad \text{with} \quad \cos q = \text{sn}(\frac{k_1\phi}{\pi}, k_1) ,$$

the corresponding energy

$$E = J_x \text{sn}(2\zeta,\ell)\,\frac{K_1}{K_\ell'}\,\sqrt{\sin^2 q + k_1'^2\cos^2 q} .$$

The momentum of the nth B-state is $p = \frac{q}{a}(\text{mod } \frac{\pi}{a})$ with

$$\cos q = [\text{sn}^2(\frac{K_1\phi}{\pi}) - \text{cn}^2(\frac{K_1\phi}{\pi})(\text{sn}'y)^2][\text{sn}^2(\frac{K_1\phi}{\pi})(\text{cn}'y)^2 + (\text{sn}'y)^2]^{-1} ,$$

the corresponding energy

$$E = J_x\, \text{sn}(2\zeta,\ell)\frac{K_1}{K_\ell'}\,\frac{2}{\text{sn}'y}\sqrt{1-(\text{cn}'y\cos\frac{q}{2})^2}\sqrt{\sin^2(\frac{q}{2}) + (k_1'\text{sn}'y\cos\frac{q}{2})^2}$$

where

$$\text{sn } u = \text{sn}(u,k_1), \quad \text{cn } u = \text{cn}(u,k_1)$$
$$\text{sn}'u = \text{sn}(u,k_1'), \quad \text{cn}'u = \text{cn}(u,k_1')$$

and

$$y = K_1' n(\frac{\pi}{\mu} - 1) .$$

(c → d) <u>The relation of the lattice variant of the massive Thirring model and the continuum theory</u>

The continuum limit of the lattice theory is taken by first letting N tend to infinity and then letting a tend to zero. In the continuum limit

$$\mu=\varepsilon, \quad k_1'=aM, \quad \ell=4(\frac{aM}{4})^{\varepsilon/\pi}, \quad \zeta=\frac{\varepsilon}{2}, \quad J_x=\frac{\varepsilon}{a\pi\sin\varepsilon} .$$

Therefore, the dispersion laws for the fermion (S-state)/ antifermion (A-state), and the nth fermion-antifermion bound state (B-state) become

$$E = \sqrt{M^2 + p^2}$$

and

$$E = \sqrt{M_n^2 + p^2}, \quad M_n = 2M\sin(\frac{n\pi}{2}(\frac{\pi}{\mu} - 1))$$

respectively in accordance with relativistic invariance.

Formally, the Thirring field operator in this limit is given by

$$\psi_1 = \frac{1}{\sqrt{2}}\{\chi_1^\dagger + \chi_2^\dagger\}, \quad \psi_2 = \frac{1}{\sqrt{2}}\{\chi_1 - \chi_2\}$$

where

$$\gamma^0 = \begin{pmatrix} 0 & 1 \\ 1 & 0 \end{pmatrix}, \quad \gamma^1 = \begin{pmatrix} 0 & 1 \\ -1 & 0 \end{pmatrix}$$

$$\chi_1(t,na) = [2aZ_2(a)]^{-\frac{1}{2}}\phi_n(t) \quad (n \text{ even})$$

$$\chi_2(t,na) = [2aZ_2(a)]^{-\frac{1}{2}}\phi_n(t) \quad (n \text{ odd})$$

$Z_2(a)$ being a cut-off dependent wave function renormalization constant.

(d → e) Relation of the massive Thirring model and the quantum sine Gordon theory

The correspondence between these two theories has been clarified by S. Coleman [12,13] . The adequate identifications are

$$\frac{g_{s.G.}}{m} \epsilon^{\mu\nu} \partial_\nu \phi = 2\pi (\bar{\psi}\gamma^\mu \psi)$$

$$g^2_{s.G.} = 4\pi \frac{m^2}{1+\frac{g}{\pi}}$$

where the current $\bar{\psi}\gamma^o\psi$ is defined in such a way that the charge $Q = \int_{-\infty}^{+\infty} dx(\bar{\psi}\gamma^o\psi)$ adopts integer values only.

Note the equality of the fermion number current and the topological current of the sine Gordon theory.

Identifying the fermion mass with the soliton mass we finally arrive at the desired expression for the masses of the sine Gordon breathers

$$M_n = 2M \sin \left(\frac{n}{16} \cdot \frac{g^2_{s.G.}/m^2}{1-\frac{1}{8\pi} g^2_{s.G.}/m^2}\right) .$$

In this way, for the first time in the history of relativistic local quantum field theory, a non-trivial mass spectrum has been calculated exactly.

In the meantime, in the infinite volume limit of the lattice variant of the massive Thirring model, M.Lüscher has constructed a denumerably infinite set of conserved, commuting "local" charges. In substance, he takes the unitarily transformed higher logarithmic derivatives of the transfer matrices of the eight-vertex models at the point $V = \zeta$

$$U \frac{\partial^n}{\partial V^n} \ln\left({\mathbb{T}^{(\pm)}}^{-1}(\zeta)\, \mathbb{T}^{(\pm)}(V)\right)\Big/_{V=\zeta} U^{-1} \quad n=1,2,3,\ldots$$

which commute among themselves (in particular with H). Subsequently, he rewrites the charges in the Fermion language and passes to the infinitely long chain. Although the continuum limit is rather delicate and has not been performed yet (i.e. the construction of infinitely many conserved commuting local charges in the massive Thirring model is still an open problem), some qualitative conclusions about the soliton, antisoliton, breather reactions in the massive Thirring model can be drawn anyway: There is no production of solitons, antisolitons or breathers of any type, the momentum sets for incoming solitons combined with antisolitons (incoming breathers of type n) and outgoing solitons combined with antisolitons (outgoing breathers of type n) are identical.

This result greatly facilitates the analysis of the scattering matrix of the fully quantized sine Gordon theory, the obvious next goal.

REFERENCES

[1] Excellent introductions to the soliton concept are: A.C. Scott, F.Y.F. Chu and D.W. McLaughlin: Proceedings of the IEEE 61, 1443 (1973)
S. Coleman: Classical Lumps and Their Quantum Descendants, School of Subnuclear Physics "Etteore Majorana", Erice 1975

[2] W.E. Sacharov and A.B. Shabat: JETP 61, 118 (1971)
L.A. Takhtadzhyan: JETP 66, 476 (1974)

[3] N.J. Zabusky and M.D. Kruskal: Phys. Rev. Lett. 15, 24o (1965)

[4] G.'t Hooft: Nucl. Physics B 79, 276 (1974)
J. Arafune, P.G.O. Freund and C.J. Goebel: J.Math. Phys. 16, 433 (1975)

[5] S. Coleman: Classical Lumps and Their Quantum Descendants, School of Subnuclear Physics "Ettore Majorana", Erice 1975

[6] L.D. Fadeev: Some Comments on the Many-Dimensional Solitons, Cern preprint TH 2188 (1976)

[7] L.D. Fadeev: Trudy Matem.Inst.im.Steklova 73, 314-336 (1964)

[8] M.A. Naimark: Dokl.Akad.Nauk SSSR, 85, 41 (1952), 89, 213 (1953) and Trudy Mosk.Matem.Obshch. 3, 181 (1954); B.Ya. Levin: Dokl. Akad.Nauk SSSR, 1o6, 187 (1956); B.S. Pavlov: Topics in Math.Physics 1,87 (1967) ed. by M.Sh. Birman

[9] L.A. Takhtadzhyan and L.D. Fadeev: Theor. and Math. Physics 21, 1o46 (1974)

[1o] V.S. Buslaev and L.D. Fadeev: Dokl. Akad. Nauk SSSR 132 (196o)

[11] A. Luther: Eigenvalue Spectrum of Interacting Massive Fermions in One Dimension, NORDITA 76/8

[12] S. Coleman: Phys. Rev. D 11, 2o88 (1975)

[13] B. Schroer and T.T. Truong: Equivalence of sine-Gordon and Thirring Model and Cumulative Mass Effects. FUB HEP 6/76
[14] M. Lüscher: Dynamical Charges in the Quantized Renormalized Massive Thirring Model DESY 76/31 Private Communications
[15] R.J. Baxter: Ann. Phys. (N.Y.) 70, 193 and 323 (197o)
[16] J.D. Johnson, S. Krinsky and B.M. McCoy: Phys. Rev. A 8, 2526 (1973)

CLASSICAL STATIC GAUGE-FIELD SOLITONS IN THREE SPACE DIMENSIONS

L. O'Raifeartaigh

DIAS, Dublin, Ireland and

IHES, Bures-sur-Yvette, France

INTRODUCTION

These three lectures will be complementary to the earlier lectures on solitons in that they will deal with solitons in physical, three-dimensional space. On the other hand, the solitons will be purely classical. The relevance of of gauge-theory to solitons solutions is that, according to an argument of Derrick[(1)], static solitons (finite energy solutions of the field equations) can be constructed from scalar fields alone in only one space dimension, and a natural possibility to overcome this problem is to use vector, or gauge fields. Recently it has been shown that static solitons in two, three and even four dimensions, can indeed be constructed in this way. We shall be interested principly in the static solitons in three dimensions, whose existence was first noted [(2)] by 't Hooft and Polyakov. In the first lecture the 't H-P soliton, together with two simple generalizations of it, will be presented. In the following two lectures we shall be concerned with further generalizations of this soliton, particularly generalizations with higher values of the magnetic charge. The interest in such solutions is to know whether excited states of the basic solitons and static bound states of two or more basic solitons, exist, and to prepare the way for the scattering theory of solitons in three dimensions. Accordingly the second lecture will be devoted to the study of the boundary conditions for the existence of solitons, on the sphere at infinity, S_2.

A powerful tool for such a study is the theory of homotopy groups and this theory will be briefly reviewed and applied.

The boundary conditions on S_2 give,however, only necessary conditions for the existence of solitons i.e. regular finite energy solutions of the field equations throughout the whole of E(3)-space inside R_2, and so in the last lecture we consider the problem of obtaining solutions throughout E(3). Here we shall be considering not the general problem of existence of solutions, but only that of the explicit construction of solutions. Our results will be that the simplest generalizations of the 'tH-P Ansatz, namely the assumption of spherical symmetry, or the assumption of radial seperation of variables, both lead back to the original 'tH-P solution. From the point of view of new solutions this result is disappointing and indicates that new explicit solutions may be difficult to construct. From the pedagogical point of view, however, the result is perhaps interesting in that provides a derivation of the 'tH-P solution from more general assumptions.

LECTURE 1

Let us begin by explaining in a little more detail why the search for static solitons in more than one space dimension leads us to introduce gauge fields i.e. let us consider Derricks argument[(1)] in a little more detail. Let

$$H = \int d^n x \left[\frac{1}{2}(\nabla\phi)^2 + V(\phi)\right], \qquad V(\phi) \geq 0 \tag{1.1}$$

be the static Hamiltonian for a single scalar field, or multiplet of scalar fields, where the positivity of $V(\phi)$ is not a real restriction so long as V is bounded below. Suppose now that we are looking for finite-energy (FE) solutions of the field equations corresponding to H. For such solutions H will be extremal

$$\left(\frac{\delta H}{\delta \phi}\right)_{\phi=\phi_o} = 0, \qquad \text{where } \phi_o\text{=solution.} \tag{1.2}$$

But now consider the special variation

$$\phi_o(x) \to \phi(x) = \phi_o(\lambda x) \tag{1.3}$$

for any real non-zero λ. Then (1.2) implies in particular that

$$\left\{\frac{\delta H}{\delta \lambda}\right\}_{\lambda=1} = 0 \tag{1.4}$$

But from (1.1) and (1.3) one easily sees that

$$H(\lambda) = \lambda^{-n+2} \int d^n x \, \frac{1}{2} \, (\nabla \phi_o)^2 + \lambda^{-n} \int d^n x \, V(\phi_o) \tag{1.5}$$

where each integral must converge seperately, since each term in (1.1) is positive. Hence

$$(-1)\left\{\frac{\delta H\,(\lambda)}{\delta \lambda}\right\}_{\lambda=1} = (n-2)\int d^n x \, \frac{1}{2}(\nabla \phi_o)^2 + n\int d^n x \, V(\phi_o) = 0 \tag{1.6}$$

But since the integrals in (1.6) are positive, it is clear that we can obtain a non-trivial solution only if n-2 is negative i.e. only if n=1. Thus no static solitons will exist for $n \geq 2$. Actually, to complete the argument one has to check that the boundary term

$$\int r^{n-1} \, d\Omega \left\{\frac{\partial \phi}{\partial r}\right\} \delta \phi$$

which is discarded in passing from the minimization of H to the field equations, really does vanish for the particular variation (1.3). Since for (1.3) $\delta\phi = \delta\lambda \; r\frac{\partial \phi}{\partial r}$ one easily sees that the vanishing of the boundary term is guaranteed by the convergence of the first integral in (1.6).

The standard way to circumvent this problem is to introduce vector, or gauge, fields, $\vec{A}(x)$ in addition to the scalar fields. This is done within the Yang-Mills framework, which requires that the Hamiltonian be invariant with respect to a gauge group G which is a direct product of compact lie groups $G(\vec{x})$ at each space-point $\vec{x}$. Then the Hamiltonian takes the form

$$H = \int d^n x \{\tfrac{1}{2}(D\Phi)^2 + V(\Phi) + \tfrac{1}{2} F^2\} \tag{1.7}$$

where Φ are the scalar fields and belong to a representation R of G, and $\vec{D}\Phi$ is the covariant derivative of Φ i.e.

$$\vec{D}\Phi = \vec{D}\Phi_a = \vec{\nabla}\Phi + e\vec{A}_\alpha\, t^\alpha_{ab}\,\Phi_b \tag{1.8}$$

where t^α are the generators of R, e being a coupling constant whose value is not fixed. $V(\Phi)$ is a G-invariant potential, and

$$\vec{F}=\vec{F}_\alpha\tau_\alpha, \text{ where } \vec{F}_\alpha = \frac{1}{e}[\vec{D},\vec{D}]_\alpha = \text{curl } \vec{A}_\alpha + e[\vec{A},\vec{A}]_\alpha$$

$$F^2 = \vec{F}_\alpha.\vec{F}_\alpha \tag{1.9}$$

and τ_α are the generators of the adjoint representation of G. The advantage in introducing the $\vec{A}$-fields, is that if we now try to repeat the Derrick argument we find that in order to have

$$D\Phi(\Lambda x) \rightarrow \lambda\ D\Phi(x) \tag{1.10}$$

as before, when we change integration variable, we must have

$$\vec{A}(\lambda x) \rightarrow \lambda\ \vec{A}(x) \tag{1.11}$$

and hence

$$\vec{F}(\lambda x) \rightarrow \lambda^2\vec{F}(x) \tag{1.12}$$

But then we have

$$(-1)\left(\frac{\partial H}{\partial\lambda}\right)_{\lambda=1} = (n-2)\int\frac{1}{2}(D\Phi)^2 + n\int V(\Phi)+(n-4)\int\frac{1}{2}F^2 \tag{1.13}$$

and this expression is not manifestly positive if $n<4$. So we have stepped up the possibilities from $n<2$ to $n<4$. In fact, even for $n=4$ (1.13) implies only that the field Φ is constant, but says nothing about $\vec{F}$, so that a soliton with pure gauge fields $\vec{F}$ for $n=4$ is not ruled out. And, indeed, such solitons (instantons) have recently been found[(3)]. However here we shall be interested only in $n=2$ and 3, particularly $n=3$.

For $n=2$ a finite energy (FE) solution for the Hamiltonian system (1.7) exists for the gauge group $G=U(1)$, in which case $\vec{A}(x)$ is just the ordinary maxwell magnetic potential[(4)]. The most recent discovery[(2)], however, is that for G=SU(2) a solution exists for n=3. The rest of this lecture will be devoted to presenting this solution in its most general known form.

Before going on to describe the solution for n=3 I first have to say a word about the potential $V(\Phi)$. The $\vec{A}(x)$ fields in (1.7) do not have explicit mass-terms, on account of the gauge invariance, and unless they acquire masses (or at least unless all but an abelian set of them acquire masses) there will be an infrared problem. On the other hand, if the $\vec{A}$-fields are simply given an ad hoc mass-term, the corresponding quantized theory will not be renormalizable. Hence the potential $V(\Phi)$ is chosen so as to provide the A-fields with masses through spontaneous symmetry breaking. That is to say $V(\Phi)$ is chosen so that its minimum occurs not at $\Phi=0$ but at $\Phi=c\neq 0$, where c is a constant vector in the representation R. If one then makes the shift $\Phi\rightarrow\Phi'=\Phi-c$ so that $\Phi'=0$ at the potential minimum, one sees that the kinetic term for Φ becomes, from (1.7),

$$\frac{1}{2}(D\Phi)^2 = \frac{1}{2}(Dc)^2 + \ldots = \frac{e^2}{2}(A_\alpha t_\alpha c)^2 + \ldots$$

$$= \frac{1}{2}(t_\alpha c, t_\beta c)\ \vec{A}_\alpha \cdot \vec{A}_\beta + \ldots \quad (1.14)$$

where the dots denote terms of at least third order in $\vec{A}$ and Φ' (interaction terms). Thus the fields $\vec{A}$ acquire the mass-squared matrix

$$M_{\alpha\beta} = (t_\alpha c, t_\beta c) \quad (1.15)$$

and only those A's which lie in the direction of the stability group H of the vector c i.e.

$$\vec{A}(n) = \vec{A}_\alpha n_\alpha \quad \text{where } (n_\alpha t_\alpha)c = 0,\ n_\alpha n_\alpha = 1 \quad (1.16)$$

remain massless. Thus all but an abelian set of $\vec{A}$'s will acquire masses if the stability group H of C is abelian. The process (1.14) by which the $\vec{A}$'s acquire masses due to the spontaneous breakdown of $V(\Phi)$ is part of what is called the Higgs mechanism[(5)]. From now on, therefore, we shall assume that $V(\Phi)$ is a potential with spontaneous breakdown (Higgs potential). In particular, if V is polynomial of at most fourth degree, then for G=SU(2), V will necessarily take the form

$$(\Phi) = \lambda(\Phi^2-c^2)^2, \tag{1.17}$$

where $\lambda>0$ is the Φ^4 coupling constant.

After this excursion into Higgs theory we return to our main problem of describing the known FE solutions of (1.7) for n=3. The basic solution[2] is obtained by putting Φ in the adjoint representation of G=SU(2) (along with $\vec{A}$, which is necessarily in the adjoint) and then making the Ansatz

$$A_i^\alpha(x) = \varepsilon_{i\alpha a}x_s\left(\frac{1-K(r)}{r^2}\right), \quad \Phi^\alpha(x)=\delta_{\alpha s}x_s\left(\frac{H(r)}{r^2}\right) \tag{1.18}$$

The Ansatz (1.18) couples space and internal indices and so requires some apologia. The point is that, since, in a gauge theory, there is a separate Lie group G(x) = $SU(2)_x$ at each point x of three-space, we are free to choose the directions in $SU(2)_x$ in any way we please for each x. We use this freedom in making the Ansatz (1.18), which is therefore restrictive but self-consistent. However, (1.18) also requires a slight adjustment of the Higgs mechanism, since if we compare $\Phi^\alpha=c_\alpha$ at minimum of V with (1.18) we see that c_α can no larger be constant, but must be of the form

$$c_\alpha = c\, x_\alpha/r, \tag{1.19}$$

where c is constant. Thus c_α is angular dependent. The angular dependence of c_α has two consequences: First, the mass-matrix (1.15) becomes angular dependent. But this will not matter, since its eigenvalues, which are the physical quantities, remain constant. Second, comparing (1.19) with (1.18) we see that we have the boundary condition

$$\underset{r=\infty}{\mathrm{Lt}}\ \Phi^\alpha = c\, x_\alpha/r \quad \text{or} \quad \underset{r=\infty}{\mathrm{Lt}}\left(\frac{H(r)}{r}\right) = c. \tag{1.20}$$

Note that the Higgs field Φ does not go to zero as $r\to\infty$. This fact will be crucial for the existence of a soliton.

Having settled these points raised by the Ansatz (1.18) we are now ready to go. First we write the field equations for the Hamiltonian (1.7), namely

$$D^2\Phi = \frac{\partial V}{\partial \Phi} \qquad \vec{D}\times\vec{F}^\alpha = e(\Phi, t_\alpha \vec{D}\Phi) \tag{1.21}$$

where the inner product means summation over the group indices. In the static case the field equations (1.21) are also the minimalization equations for H. We then insert the Ansatz (1.18) in (1.21). We then find that the indicial part of (1.21) is identically satisfied (a result which we shall see later is not trivial) and that we are left with the two radial equations

$$r^2H'' = 2HK^2 + r^2H\ W(H/cr)$$
$$r^2K'' = K(K^2-1) + K\ H^2 \tag{1.22}$$

where the function $W(H/cr)$ depends on the precise form of the potential $V(\Phi)$ in (1.17). For example (1.17) leads to

$$W(H/rc) = \lambda\left[(H/rc)^2-1\right] \tag{1.23}$$

One is looking for regular finite-energy solutions of (1.22) subject to certain boundary conditions. Finite energy means that when the solutions are re-inserted in the Hamiltonian (1.7) the integral converges. The boundary conditions are as follows: regularity at $r=0$ requires that

$$1-K(o) = H(o) = 0 \tag{1.24}$$

and indeed we shall find also

$$K'(0) = H'(0) = 0 \tag{1.25}$$

On the other hand, at $r\to\infty$ we demand

$$K(r)\to o \qquad H(r)\to rc \tag{1.26}$$

The first condition in (1.26) corresponds to $\vec{A}(x)\to 0$ as $r\to\infty$ and the second to $\Phi_\alpha(x)\to c_\alpha$, as discussed earlier.

Using trial functions, 't Hooft[(2)] was able to show heuristically that the coupled non-linear equations (1.22) with the boundary conditions (1.24) (1.26) have a finite energy solution. This result was subsequently established rigorously[(16)], and thus for $G=SU(2)$ the Hamiltonian (1.7) has soliton solutions.

For the special case $\lambda = 0$ (zero Higgs

potential) the solution actually takes a closed form in terms of elementary functions, namely,

$$K(r) = \frac{rc}{\sin hrc} \qquad H(r) = rc\,\frac{\cos hrc}{\sin hrc} - 1 \qquad (1.27)$$

Notice that in the special case $\lambda=0$ the boundary condition $H(r)\to cr$ is ad hoc, since it is not required by the potential. The physical meaning of $\lambda=0$, $c\neq 0$ is that the mass of the Higgs field is zero but the masses of the gauge fields are not (except for the A in the direction of the stability group).

Can one generalize the solutions found above? The next two lectures will be concerned with this question, and for the remainder of this lecture, we shall consider only two fairly simple generalizations. The first generalization(7) is to add a static electric potential to the static magnetic potential $\vec{A}$, by adding terms

$$-\frac{1}{2}(\vec{D}A_o)^2 - \frac{1}{2}(A_o\Phi, A_o\Phi) \qquad (1.28)$$

to the Lagrangian density, and making the Ansatz

$$A_o^\alpha(x) = \delta_{\alpha r}\, x_r (J(r)/r^2) \qquad (1.29)$$

Then the second term in (1.28) drops out because A_o and Φ are parallel in the group, the indicial equations for the first term are identically satisfied, and the radial equations become

$$r^2K'' = K(K^2-1) + K(H^2-J^2)$$

$$r^2H'' = 2HK^2 + r^2H\, W(H/rc),$$

$$r^2J'' = K^2J. \qquad (1.30)$$

It is not difficult to see that if solutions to the original radial equations exist, so do solutions to (1.30). In particular for $\lambda=0$ it is easy to see that the solution is

$$K(r) = \frac{rc}{\sinh rc} \qquad \frac{H(r)}{\cosh\gamma} = \frac{J(r)}{\sinh\gamma} = rc\,\frac{\cosh rc}{\sinh rc} - 1 \quad (1.31)$$

where γ is an arbitrary parameter.

A second generalization is to let the Higgs field belong to any other integral representation of G=SU(2) with isopin t(t+1) say, and to generalize the Ansatz (1.18) to

$$\Phi^a(x) = c\sqrt{\frac{4\pi}{2t+1}}\, y_a^t(\Omega)\,[H(r)/r], \quad a=1\ldots 2t+1 \qquad (1.32)$$

where y(Ω) are the spherical harmics (in a real basis). Eqn. (1.32) equates the space-spin of Φ(x) to the isospin t, and thus generalizes (1.18) from ℓ=t=1 to ℓ=t=any positive integer. Note that the generalization (1.22) cannot be made for the gauge-fields $\vec{A}$ and A_o since the gauge fields necessarily lie in the adjoint representation.

Inserting (1.32) into the equation of motion (1.21) one finds that the indicial equations are identically satisfied once again and that the radial equations become

$$r^2K'' = K(K^2-1)+K\left[\frac{\ell(\ell+1)}{2}H^2-J^2\right]$$

$$r^2H'' = \ell(\ell+1)K^2H + r^2H\,W(H|rc) \qquad (1.33)$$

$$r^2J'' = 2K^2J$$

The existence proof[16] for solutions of (1.22) can be generalized immediately to (1.33) so that these equations also have solutions. The main interest of the generalization (1.32 is that it provides a mass-formula according to the equation

$$M(t) = T_{oo} = \frac{4\pi}{e^2}\int_o^\infty \frac{dr}{r^2}\,\{(rK')^2+\tfrac{1}{2}(rH'-H)^2+\tfrac{1}{2}(rJ'-J)^2$$

$$+\tfrac{1}{2}(K^2-1)^2+\tfrac{1}{2}K^2J^2+\frac{t(t+1)}{4}K^2H^2 + cr\,U(\tfrac{H}{cr})\} \qquad (1.34)$$

where $U'(X) = W(X)$, $U(O) = O$. It can be shown that M(t) actually increases slowly but monotonically with t and tends to a finite limit as $t\to\infty$.

Finally we come to the question of interpreting the solutions. For SU(2) there are only two possible stability groups, H=1 and H=U(1), and our Ansätze for Φ (integer representations with $\ell=t$) correspond to the non-trivial case H=U(1). Let (,) and $\wedge$ denote scalar and cross product in the group space respectively. Then the Higgs theory tells us that we have one massless gauge field (u,A) in the direction u of the stability group U(1), and two massive gauge fields $u\wedge A$ orthogonal to u. For $\lambda\neq 0$ the Higgs field also is massive. The gauge field with zero mass is identified with the electromagnetic field and so u is identified with the electromagnetic direction in the adjoint representation.

The new feature of the model is that, since $\phi(\Omega)$ is Ω-dependent, the direction u of its stability group u is Ω-dependent, $u=u(\Omega)$. Since u then does not compute with derivation, we have a problem in defining the magnetic field $\vec{B}$, which for constant u, u_o say, is just the curl of the magnetic potential

$$\vec{B}_o = \vec{\nabla}\times(u_o,\vec{A}) = (u_o,\vec{\nabla}\times\vec{A}) \qquad (1.35)$$

The problem is to find a covariant quantity which reduces to (1.35) when $u=u_o$=constant. By covariant here is meant a quantity which depends only on u, $\vec{D}$ and $\vec{F}$. This problem was solved by 't Hooft and Polyakov[(2)] who defined $\vec{B}$ to be

$$\vec{B} = (u,\vec{F}) + \frac{1}{2e}(u,\vec{D}u\wedge\vec{D}u) \qquad (1.36)$$

The quantity (1.36) is manifestly covariant, and if we calculate it out directly we find that it is

$$\vec{B} = \vec{\nabla}\times(u,\vec{A}) + \frac{1}{2e}(u,\ \vec{\nabla}u\wedge\vec{\nabla}u) \qquad (1.37)$$

in which form it manifestly reduces to (1.35) for $u=u_o$= constant. The most important feature of (1.37) is that $\vec{B}$ is not a pure curl as in maxwell theory, but has an extra term. on the other hand, this extra term is independent of the magnetic potential $\vec{A}$ and depends only on EM direction vector u. In particular if we take the divergence of $\vec{B}$, the curl term drops out, and we obtain

$$\nabla\cdot B = \frac{1}{2e}(\vec{\nabla}u;\vec{\nabla}u\ \vec{\nabla}u) \qquad (1.38)$$

$$= \frac{1}{2e}\,\varepsilon_{\alpha\beta\gamma}\,\varepsilon_{ijk}\,\nabla_i U^\alpha \nabla_j U^\beta \nabla_k U^\gamma$$

If we now integrate (1.38) over all space we obtain

$$M = \int d^3x \quad \nabla\cdot\vec{B} = \frac{1}{2e}\int d\Omega\,(u,\ \vec{\partial}u_\wedge\vec{\partial}u\ ) \qquad \vec{\partial}=r\vec{\nabla} \tag{1.39}$$

so that in general there is a magnetic charge M. In particular, for the 't-Hooft-Polyakov model (and the generalizations discussed above)

$$u(\Omega) = u^\alpha(\Omega) = \phi^\alpha(\Omega) = x_\alpha/r \tag{1.40}$$

and hence by a straightforward calculation from (1.39),

$$M = \frac{4\pi}{e}\ . \tag{1.41}$$

Thus for this model there is a magnetic charge and it is quantized in units of $4\pi/e$ as in the Dirac[8] string theory of magnetic monopoles. Thus the mass-zero part of the gauge field describes a magnetic monopole.

The generalization to include A_o is trivial, because, on account of the Ansatz (1.18) (1.29) we have $(u,A_o)=A_o$ and so there are no new features. In particular the electric field is simply

$$\vec{E} = \vec{\nabla}A_o = \frac{1}{e}\,\vec{\nabla}\left(\frac{J(r)}{r}\right) = \vec{r}\,\frac{1}{r}\,\left(\frac{J}{r}\right)'\,\frac{1}{e} \tag{1.42}$$

and the electric charge Q is

$$eQ = \int d^3x\,\vec{\nabla}\cdot E = \int d\Omega\,\left[r^2\left(\frac{J}{r}\right)'\right]_{r=\infty} = 4\pi\left[rJ'-J\right]_\infty \tag{1.43}$$

For the exact, $\lambda=0$, solutions of the field equations we have

$$Q = \frac{4\pi}{e}\sinh\gamma \tag{1.44}$$

Thus the mass-zero part of the gauge field in this case describes a soliton of electric charge $4\pi e^{-1}\sinh\gamma$ and magnetic charge $4\pi e^{-1}$. Such a system, with non-zero electric and magnetic charge, is called a dyon.

I should like to conclude this lecture by emphasizing the following point: the existence of the 'tH-P soliton is possible only because of the boundary condition

$(\Phi,\Phi) \rightarrow$ constant $\neq 0$ as $r \rightarrow \infty$, and this boundary condition in turn, is only possible because the potential is a Higgs potential. (Notice that the exact solution (1.27) collapses if c=0). This is actually a general result. Thus the Higgs potential, which is originally introduced in order to provide the gauge fields with masses without destroying the renormalizability plays a crucial role in the construction of soliton solutions.

LECTURE 2

In this lecture we shall obtain two further insights into the nature of the magnetic charge M found in lecture 1. These insights will not only be of interest in themselves, but the second of them will provide a powerful tool for classifying other possible solutions.

The first observation[9] is that if we define the current

$$k_{\mu} = \varepsilon_{\mu\lambda\sigma\nu}\, \partial_{\sigma}\, F^{EM}_{\nu\lambda} \tag{2.1}$$

which would be identically zero in maxwell theory, we find that the magnetic charge M defined in the last lecture is given by

$$M = \int k_{o}(x)\; d^{3}x \tag{2.2}$$

i.e. the magnetic charge is just the charge associated in the usual way with a current. Further since k_{μ} is identically conserved on account of the anti-symmetry of the Levi-Civitá symbol, that is,

$$\partial_{\mu} k_{\mu} \equiv 0 \tag{2.3}$$

M is identically conserved

$$\dot{M} \equiv 0 \tag{2.4}$$

Nevertheless M is not a conventional charge, since it does not generate any symmetry transformation. In fact, M does not generate any transformations at all, since it commutes with all the field variables. To see this we note from (1.39) that M contains only the space-derivatives of Boson fields. Hence its equal-time commutators with all the fields is zero. But M is also identically conserved. Hence it commutes with the field variables at all times. Thus M lies in the centre of the algebra of fields and is a superselection operator. Furthermore, M is not an arbitrary superselection operator but is constructed out of the fields themselves. Thus it is a Casimir operator of the algebra of fields.

Of course, the word 'commutes' in the above discussion must be taken cum grano salis since we do not have here a quantum theory. However the word commutation is meant in the sense of Poisson brackets for the classical theory, and it is not unreasonable to expect that, once a quantized theory has been constructed, the result will hold in the sense of ordinary commutation in that theory.

For the second observation[10] concerning the nature of M we need some concepts and results from homotopy theory[11] and the next few paragraphs will be devoted to a resumé of the necessary information. We begin with the definition of the homotopy groups π_n.

Consider the unit circle S_1, parametrized in the usual way with ϕ where $0 \leq \phi \leq 2\pi$, and a mapping $f(\phi)$ from S_1 into any compact topological space S, such that $f(\phi)$ is (i) continuous, and (ii) single-valued ($f(2\pi)=f(0)$). For example, S could be the doughnut of Fig.1, with the closed curves on it as values of f.

Two such mappings $f(\phi)$ and $g(\phi)$ are called equivalent if they can be continuously deformed into one another i.e. if there exists a set of intermediate

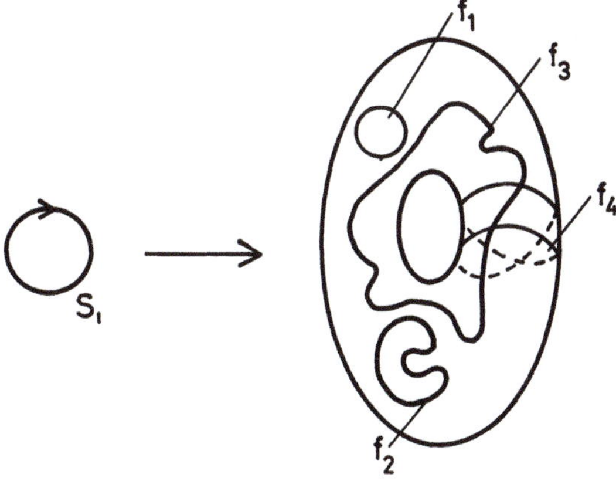

Figure 1

mappings $f(\phi,t)$ (always from S_1 into S), continuous in t and such $t(\phi,0)=f(\phi)$ and $f(\phi,1) = g(\phi)$. Clearly the mappings f_2 and f_3 in Fig. 1 are inequivalent, but the mappings f_1 and f_2 are equivalent. Let $e=\{f\}$ be the equivalence class of any map f i.e. the set of all mappings equivalent to f.

Next, on the space of mappings define an operation of multiplication $\otimes$ by the process of cutting curves and joining the loose ends together to obtain new closed curves, as in Fig. 2, for example

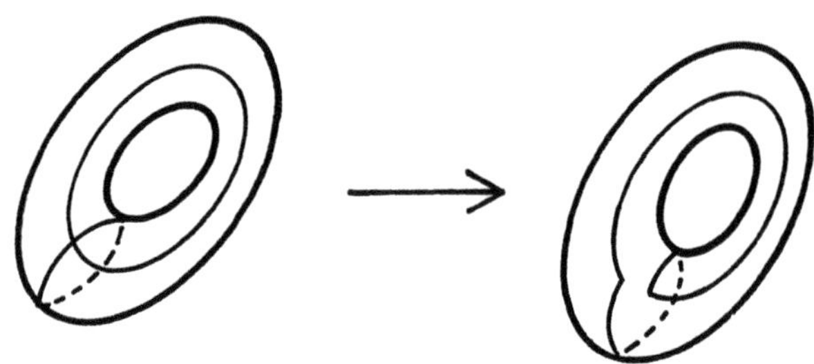

Figure 2

The formal definition of such multiplication is

$$f = f_1 \otimes f_2 \text{ if } \quad f(\phi) = f_1(2\phi) \text{ for } 0\leq\phi\leq\pi,$$

$$f(\phi) = f_2(2\phi) \text{ for } \pi\leq\phi\leq 2\pi.$$

Under the operation $\otimes$ the equivalence classes $e=\{f\}$ of the mappings f form a group. The group formed by the equivalence classes is called the first homotophy group S and is denoted by $\pi_1(S)$.

In the same way we can consider the set of all continuous, single-valued, mappings $f(\Omega)$ of the surface S_2 of the sphere in 3-dimensions into S, where Ω are the polar angles $\Omega=(\Theta,\phi)$, construct the equivalence classes $e = \{f(\Omega)\}$, define an operation of multiplication by cutting and joining, and so form a group with elements $e = \{f(\Omega)\}$. This group of equivalence classes of mappings from S_2 into S is called the second homotopy group of S and is denoted by $\pi_2(S)$. And in general it is clear that for $S_n \to S$, where S_n is the sphere in n+1 dimensions, we have $\pi_n(S)$. Let us now consider some relevant examples:

Example 1: $\pi_1(S_1)$. This case corresponds to the mappings $S_1 \to S_1$. Let us parametrize the first S_1 by ϕ as usual, and the second S_1 by a similar variable χ i.e. $f(\phi) = \exp(i\chi(\phi))$. Then the single-valuedness condition $f(o) = f(2\pi)$ implies that $\chi(2\pi) = \chi(o)+2n\pi$, and geometrically it is easy to see two χ's will be equivalent if, and only if, they have the same integer n i.e. if, as ϕ goes from O to 2π, the two χ's go round the second circle the same number of times. (n is sometimes called the winding number). Hence in this case the elements e of $\pi_1(s)$ are characterized by the integers n, $\{e\}=e_n$, $n=o, \pm 1, \pm 2, \ldots$. Furthermore it is clear that for each class there is a simple representative element $\chi(\phi) = n\phi$ or $f(\phi) = \exp(in\phi)$. Finally, one sees trivially that for any χ its class-number is given by the integral formula

$$n = \frac{\int d\chi(\phi)}{\int d\phi} \tag{2.5}$$

where both integrals are to be calculated over the range $O \leq \phi \leq 2\pi$

Example 2.: $\pi_1(S_2) = O$. This just the statement that any closed curve on the surface of the sphere in three dimensions can be continuously shrunk to zero.

Example 3.: $\pi_2(S_2)$. Just as ϕ, $O \leq \phi \leq 2\pi$ is a natural parameter for S_1, the polar angles $\Omega=(\Theta,\phi)$ are natural parameters for S_2. Since the mappings we are considering here are $S_2 \to S_2$ we parametrize the first S_2 by Ω and the second by polar angles $(\omega(\Omega),\chi(\Omega))$. Then any mapping $f(\Omega)$ takes the form

$$f(\Omega) = (\sin\omega \cos\chi, \sin\omega \sin\chi, \cos\omega). \tag{2.6}$$

Since f maps the two parameters (Θ,ϕ) into the two parameters (ω,χ) one might expect that the elements of $\pi_2(S_2)$ would be characterized by two integers. Such, however is not the case, and in fact one can see geometrically that they are characterized by a single integer n, which is just the number of times the second S_2 is covered, when the first S_2 is covered once. The number n, which is sometimes called the wrapping number, clearly has the integral form

$$n = \frac{\int d(\cos\omega(\Omega))d\chi(\Omega)}{\int d\Omega} \qquad (2.7)$$

where both integrals are calculated over $0 \leq \theta \leq \pi$, $0 \leq \phi \leq 2\pi$. The elements of $\pi_2(S_2)$ are therefore e_n, $n=0,\pm 1,\pm 2,\ldots$ and a representative element for the class e_n is

$$f(\Omega) = (\sin\theta \cos n\phi, \sin\theta \sin n\phi, \cos\theta). \qquad (2.8)$$

Example 4: $\pi_n(G)$, where G is a connected compact Lie group. Here, by a mapping $S_n \to G$, we mean letting the parameters of G be single-valued functions of the parameters of S_n. The parameter space of G is a natural space into which to map S_n, because for a compact connected Lie group, the canonical parameters have ranges $[0,\pi]$, $[0,2\pi]$ and $[0,4\pi]$. A special result for Lie groups, for which the proof[11] will not be given is:

Lemma: $\pi_2(\tilde{G}_s/H) = \pi_1(H)$ where $\tilde{G}_s$ is a simply connected semi-simple compact Lie group and H is a closed subgroup of $\tilde{G}_s$, not necessarily simple connected. By taking H=1 and $H=\tilde{G}_s$ respectively, we see at once that, in particular,

$$\pi_2(\tilde{G}_s) = \pi_1(\tilde{G}_s) = 0 \qquad (2.9)$$

Similarily by taking $G_s=SU(2)$, $H=U(1)$ and noting from ordinary angular momentum theory that $SU(2)/U(1)=S_2$ we obtain from the lemma

$$\pi_2(S_2) = \pi_1(S_1) \qquad (2.10)$$

which gives an alternative explanation as to why $\pi_2(S_2)$ and $\pi_1(S_1)$ have the same parametrization. The above lemma is of great practical value, since (as we shall see) it reduces the study of π_2 in the cases of interest to that of $\pi_1(H)$, and since H is compact, and $\pi_1(G_s)=0$ the study then reduces to that of $\pi_1(C)$ where C is the abelian centre of H i.e. $C = Z \otimes U(1)^m$ where Z is discrete and $U(1)^m$ is a direct product of m U(1)'s.

After this excursion into homotopy theory we return to physics and explain why the homotopy is relevant. For this purpose we recall from lecture 1 that if the EM direction $u(\Omega)$ in the adjoint representation of the gauge group is not constant, the covariant definition of the magnetic field is

$$\begin{aligned} \vec{B} &= (u,\vec{F}) + \frac{1}{2e}(u,\vec{D}u\ \vec{D}u) \\ &= \vec{\nabla}\times(u,\vec{A}) + \frac{1}{2e}(u,\vec{\nabla}u\ \vec{\nabla}u) \end{aligned} \tag{2.11}$$

Now since $u(\Omega)$ is, by definition, a unit vector in the space of the adjoint representation, we have for SU(2)

$$u^\alpha(\Omega)u^\alpha(\Omega) = 1 \qquad \alpha=1,2,3. \tag{2.12}$$

Thus $u(\Omega)$ is a mapping from the S_2 in ordinary space to the S_2 in the space of the adjoint representation of an internal SU(2),

$$S_2^L \xrightarrow{u(\Omega)} S_2^I \tag{2.13}$$

Furthermore, $u(\Omega)$ is continuous and single-valued. It therefore belongs to one of the equivalence class e of $\pi_2(S_2)$. But, if we use (2.11) to calculate the magnetic charge, as we did in lecture 1, we obtain

$$eM = e\int d^3x \nabla\cdot B = \int d\Omega\ \frac{1}{2}\ (u,\vec{\nabla}u\ \vec{\nabla}u)\cdot\vec{n}, \tag{2.13}$$

where $\vec{n}=\vec{x}/r$, or, in component notation

$$eM = \frac{1}{2}\ \varepsilon_{ijk}\ \varepsilon_{\alpha\beta\gamma}\int d\Omega\ n_k u^\alpha \partial_i u^\beta \partial_j u^\gamma \tag{2.14}$$

But now if we write $u(\Omega)$ in the canonical form (2.6), we find after a short computation that

$$eM = \int d\Omega \quad n_k\ \varepsilon_{kij}\ \partial_i(\cos\ \omega)\partial_j\chi \tag{2.15}$$

and one easily sees that (2.15), in turn, is just

$$eM = \int d\Omega \frac{\partial(\cos\ \omega,\chi)}{(\cos\ \Theta,\phi)} = \int d(\cos\ \omega)d\chi \tag{2.16}$$

Hence finally we find that

$$\frac{eM}{4\pi} = \frac{\int d(\cos\,\omega)d\chi}{\int d(\cos\,\Theta)d\phi}\ , \tag{2.17}$$

But if we compare (2.17) with (2.7) we see that we have the remarkable result that $eM/4\pi$ is just that wrapping number for the mapping $u(\Omega)$. This is the second interpretation of M which we were seeking. In particular (2.17) implies that $eM/4\pi$ is an integer, or

$$M = \frac{4\pi}{e} N, \quad N=0,\pm1,\pm2,... \tag{2.18}$$

Thus the single-valuedness of the mapping $u(\Omega)$ quantizes M in units of $4\pi/e$, as in the Dirac string monopole. In the particular case of the 't Hooft-Polyakov soliton (and its generalizations of lecture 1) the mapping $u(\Omega)$ is the unit map from S_2^L to S_2^I,

$$u^\alpha(\Omega) = \delta_{\alpha a}\, n_a, \quad n_a = x_a/r \tag{2.19}$$

and hence, N=1, or $eM/4\pi = 1$ as we have found before.

The result (2.18) immediately raises the question. Can one construct solutions with $|N|>1$? It turns out that such solutions (if they exist) may not be easy to construct. To see this, we recall that the mappings $u(\Omega)$ which define the EM direction, fall into equivalence classes $\{u(\Omega)\}_N$ with representative elements

$$u_N(\Omega) = (\sin\theta\cos N\phi,\ \sin\theta\sin N\phi,\ \cos\theta) \tag{2.20}$$

and that any $u(\Omega)$ in a given class can be continuously deformed into $u_N(\Omega)$. Actually such continuous deformations can be implemented by gauge transformations, so that the gauge transformations also can be d vided into equivalence classes . The problem, therefore, is to find finite-energy solutions of the Hamiltonian system (1.7) such that the EM direction will be defined by (2.20) with $|N|>1$. The difficulty in constructing such a solution is that the obvious Ansatz

$$\Phi(r,\Omega) = u_N(\Omega)\left(\frac{H(r)}{r}\right) \tag{2.21}$$

will not work. With this Ansatz the indicial equations are <u>not</u> automatically satisfied. Hence the best one can do for $|N|>1$ is to demand the boundary condition

$$\underset{r=\infty}{\mathrm{Lt}}\ \Phi\ (r,\Omega) \to u_N(\Omega) \tag{2.22}$$

Eq. (2.27) is a Dirichlet boundary condition for the fields inside S_2 i.e. throughout E(3), and the question then is whether solutions with this boundary condition exist (and are unique).

So far we have discussed only SU(2), and, in searching for $|N|>1$ we should also consider other gauge groups. For this purpose we return to the Hamiltonian (1.7) for general compact G, and consider the conditions that are placed on the fields on S_2 by the finiteness of the energy. Since the convergence of the integral requires that $r^3 H(x) \to 0$ as $r \to \infty$ we see that (subject to suitable smoothness conditions) the finite-energy requires that

$$\underset{r=\infty}{\mathrm{Lt}}\ \Phi(r,\Omega) = \phi(\Omega), \qquad \underset{r=\infty}{\mathrm{Lt}}\ r\vec{A}(r,\Omega) = \vec{a}(\Omega) \tag{2.23}$$

and that

$$d\phi \equiv \partial\phi + ea_{\wedge}\phi = 0, \qquad \vec{\partial} \equiv r\vec{\nabla} \tag{2.24}$$

$$\text{where } a_{\wedge}\phi \equiv a_\alpha t_\alpha \phi$$

Equation (2.24) will be referred to henceforth as the finite-energy (FE) condition, eq. (2.23) being understood. Eq. (2.24) states that on S_2 the convariant derivative of $\phi(\Omega)$ vanishes. One immediate consequence of (2.24) is that

$$(\phi(\Omega),\phi(\Omega)) = \text{constant}, \tag{2.25}$$

a result that is also given by the Higgs potential, which adds the information that the constant is not zero (otherwise there is no spontaneous symmetry breaking). Eq. (2.25) shows that $\phi(\Omega)$ maps S_2 into S_{r-1} where r is the dimension of the adjoint representation of G (and hence in the semi-simple case is also the order of G). Thus the generalization of $\pi_2(S_2)$ for SU(2) is $\pi_2(S_{r-1})$ for G. In practice, however, it is not so convenient to study $\pi_2(S_{r-1})$ directly, but to proceed as follows:[10]

A second consequence of the FE condition (2.24) is that

$$\phi(\Omega) = U(g(\Omega))\phi(0) \qquad (2.26)$$

where U(g) is the representation of G to which ϕ belongs, and $g(\Omega)$ is an element of G, whose Ω-dependence is obtained by integrating (2.24) and hence is very complicated. Fortunately the form of $g(\Omega)$ will not be needed. The only information we need from (2.26) is that $\phi(\Omega)$ defines a continuous single-valued map $g(\Omega)$ from S_2 into the group G itself. More precisely, since

$$U(g(\Omega))U(h)\phi(0) = U(g(\Omega))\phi(0) \qquad (2.27)$$

where h is any element of the stability subgroup H of $\phi(0)$, $g(\Omega)$ is a continuous single-valued map from S_2 into G/H. Hence the study of $\pi_2(S_{r-1})$ can be replaced by the study of $\pi_2(G/H)$. The great advantage of this is that if G is simply connected and semi-simple, as it is in most cases of interest, the lemma mentioned above then reduces the study to that of $\pi_1(H)$. Thus the superselection Casimirs M are determined by the stability group[(12)] H of $\phi(0)$.

For G=SU(2) the only non-trivial stability group is U(1), and we have already seen that $\pi_1(U(1))=\pi_1(S_1)=e_n$, n=o,±1,±2,... . For SU(3) if ϕ is in the 3-dimensional quark representation, the stability group is SU(2), and since $\pi_1(SU(2)) = 0$ as we saw above, there no FE solutions. For ϕ in the octet representation of SU(3), there are two possible stability groups, namely, H=U(2) and H=U(1)⊗U(1) (the latter being excluded if the Higgs potential contains the SU(3)-invariant tri-linear D-coupling). For H=U(2) we have $\pi_1(H)=\pi_1(U(1))=\pi_1(S_1)$ and so we have one Casimir H with eigenvalues $\frac{4}{e}n$, n=0,±1,±2,..., while for H=U(1)⊗U(1) we have $\pi_1(H)=\pi_1(S_1 \otimes S_1)$ and we have <u>two</u> such Casimirs. A complete classification for all compact simple Lie groups has been given in ref. 13.

Of course, just as in the case of G=SU(2), a given map $\phi(\Omega)$ or $g(\Omega)$ corresponding to an eigenvalue of M only determines a boundary condition on S_2, and one still has to consider the question as to whether FE solutions to the Hamiltonian system (1.7) with this boundary condition, exist throughout E(3).

LECTURE 3

In this lecture we establish two uniqueness results for the 't Hooft-Polyakov solution. These results indicate that explicit solutions in E(3) of magnetic charge $|N|>1$ may be much more complicated than those for $|N|=1$. But first we wish to mention some further properties of the FE condition (2.24) and tie up some of the loose ends of lecture 2. The first property is obtained by taking the inner-product of (2.24) with $t_\alpha \phi$, for any $\alpha=1...r$. One obtains at once

$$(t_\alpha \phi, \partial_i \phi) + e(t_\alpha \phi, t_\beta \phi) a_i^\beta = 0 \tag{3.1}$$

and if we recall from (1.15) the form of the gauge-field mass-matrix $M_{\alpha\beta}$ we see that (3.1) can be written in the form

$$M_{\alpha\beta} a_i^\beta = ec^2 (\phi, t_\alpha \partial_i \phi) \tag{3.2}$$

The right hand side of (3.2) is just the Higgs current, which would reduce to the ordinary electric current due to the Higgs field for G=U(1). Hence (3.2) shows that on S_2 the massive gauge-fields are completely determined by the mass-matrix and the Higgs current. Thus the boundary conditions for the massive $a(\Omega)$ on S_2 are completely determined by the boundary conditions for $\phi(\Omega)$.

The second property of the FE equation which we wish to consider is obtained from the integrability condition. Since the ∂_i satisfy the identity

$$\partial_i \partial_j - \partial_j \partial_i = r\nabla_i r\nabla_j - r\nabla_j r\nabla_i = r_i \nabla_j - r_j \nabla_i = \varepsilon_{ijk} L_k \tag{3.3}$$

eq. (2.24) will have a self-consistent solution if, and only if this identity is true on $\phi(\Omega)$ as a consequence of (2.24). Inserting, we obtain by direct computation

$$(f_{ij}^\alpha t_\alpha)\phi(\Omega) = 0 \text{ where } f_{ij} = \partial_i a_j - \partial_j a_i +$$

$$+ x_i a_j - x_j a_i + e\,[a_i, a_j] \tag{3.4}$$

The field f_{ij} is the analogue of the physical field $F_{\mu\nu}$ in Minkowski space. It is physical in the sense that if $f_{ij}=0$ then the $a(\Omega)$ field can be gauged to zero.

The immediate use of (3.4) is that from f_{ij} we can form the <u>scalar</u> field

$$f^{\alpha}(\Omega) = \left(\frac{x_k}{r}\right) \varepsilon_{kij} \; f^{\alpha}_{ij}(\Omega) \tag{3.6}$$

which then lies in the algebra of the adjoint representation, and according to (3.4) in the stability subalgebra of $\phi(\Omega)$. Hence $f(\Omega)$ is a natural candidate to define the EM direction i.e.

$$U^{\alpha}(\Omega) = f^{\alpha}(\Omega)/(f(\Omega),f(\Omega))^{1/2} \tag{3.7}$$

If the stability subgroup of $\phi(\Omega)$ is U(1), so that the EM direction is unique, then $f(\Omega)$ defines the EM direction uniquely (except in the exceptional case when $f(\Omega)$ is identically zero). Thus the integrability condition for the FE condition provides a natural EM direction independent of the representation to which $\phi(\Omega)$ belongs.

After this digression into the properties of the FE energy equation, we return to the problem of obtaining solitons with $|N|>1$. That the equations will be more complicated than in the $|N|=1$ case is indicated by the following two results:

Result 1. Spherically Symmetric[14] solitons have $|N|=1$ for G=SU(2)

Result 2. Radially seperated[15] solitons have $|N|=1$ for all compact gauge groups G

Note that these results establish the uniqueness of the 't Hooft-Polyakov solution under the conditions stated.

In the rest of this lecture we define these notions more precisely, and summarize the proofs.

Result 1. By spherical symmetry is meant[13] the existence of an auxiliary gauge field $\vec{\Lambda}(x)$ which can transform away the Ω-dependence of covariant tensors for all $x \in E(3)$. That is, if $\vec{L}$ is the ordinary angular momentum operator and $\vec{\mathcal{L}}$ denotes $\vec{L}+e\vec{\Lambda}$ then

$$\mathcal{L}_i S = 0, \qquad \mathcal{L}_i S_j = \varepsilon_{ijk} S_k, \tag{3.8}$$

where S and S_i are any covariant scalars and vectors respectively.

The condition (3.8) cannot be applied directly to the gauge field because $\vec{A}(x)$ is not covariant, but it can be applied to the covariant derivative $\vec{\nabla}+e\vec{A}$ and then yields the condition

$$\mathcal{L}_i A_j = \varepsilon_{ijk} A_k + \nabla_j \Lambda_i, \tag{3.9}$$

for $\vec{A}$. Eq. (3.9) is similar to (3.8) but has the extra term $\nabla_j \Lambda_i$ on the right hand side. Eq. (3.8) can also be applied to the Yang-Mills field

$$F_i = \frac{1}{2}\varepsilon_{ijk} \{\nabla_j A_k - \nabla_k A_j + e[A_j, A_k]\}, \tag{3.10}$$

to yield

$$\mathcal{L}_i F_j = \varepsilon_{ijk} F_k, \tag{3.11}$$

and it is perhaps worth noting that (3.11) follows from (3.9) and (3.10) by direct computation. The magnetic charge defined in (2.13) can be written as

$$M = \frac{1}{2e} \int d\Omega (\vec{L}\varepsilon ; \varepsilon_\Lambda \vec{\partial}\varepsilon) \tag{3.12}$$

where $\varepsilon(\Omega)$ is the unit vector in the EM direction in SU(2) space. Since (3.8) applies also to ε we have

$$\vec{L}\varepsilon + \vec{\lambda}_\Lambda \varepsilon = 0, \quad \text{where } \vec{\lambda}(\Omega) = \underset{r=\infty}{\mathrm{Lt}} \ \vec{\Lambda}(r,\Omega), \tag{3.13}$$

and we assume the limit exists. Taking the cross-product of (3.13) with $\vec{n}$, where $\vec{n} = \vec{r}/r$, we have the similar equation

$$\vec{\partial}\varepsilon + \vec{\mu}_\Lambda \varepsilon = 0 \quad \text{where} \quad \vec{\partial} = \vec{n} \times \vec{L} \quad \text{and} \quad \vec{\mu} = \vec{n} \times \vec{\lambda} \tag{3.14}$$

From (3.14) we see that the quantity $\varepsilon_\Lambda \vec{\partial}\varepsilon$ occurring in M is just

$$\varepsilon_\Lambda \vec{\partial}\varepsilon = \varepsilon_\Lambda(\varepsilon_\Lambda \vec{\mu}) = -\vec{\mu}+\varepsilon(\varepsilon,\vec{\mu}) . \tag{3.15}$$

But since the norm of ε is constant, $(\vec{L}\varepsilon,\varepsilon)$ is zero. Hence from (3.15) the integrand in M reduces to

$$(\vec{L}\varepsilon;\varepsilon_\Lambda \vec{\partial}\varepsilon) = -(\vec{L}\varepsilon,\vec{\mu}) \tag{3.16}$$

Using partial integration and the fact that for single-valued functions the total derivative with respect to $\vec{L}$ (though not with respect to $\vec{\partial}$) vanishes at the boundary, we then see that M itself reduces to

$$M = \frac{1}{2e}\int d\Omega(\varepsilon,\vec{L}\cdot\vec{\mu}) . \tag{3.17}$$

But now since $\vec{\mu}$ is $\vec{n}\times\vec{\lambda}$ we have the identiy

$$\vec{L}\cdot\vec{\mu} + \vec{\partial}\cdot\vec{\lambda} = 2\vec{n}\cdot\vec{\lambda} \tag{3.18}$$

Hence

$$M = \int d\Omega(\varepsilon,\vec{n}\cdot\vec{\lambda}) + \frac{1}{2}\int d\Omega(\varepsilon,\vec{\partial}\cdot\vec{\lambda}) . \tag{3.19}$$

Up to this point we have used only eq. (3.13). To evaluate the second integral in (3.19), however, we have to return to (3.9), or rather the contracted form

$$\vec{\nabla}\cdot\vec{\Lambda} = \vec{\mathcal{L}}\cdot\vec{A} \tag{3.20}$$

of (3.9). Taking the limit of (3.20) as $r\to\infty$ and then taking the inner product with ε we obtain

$$(\varepsilon,\vec{\partial}\cdot\vec{\lambda}) = (\varepsilon,\vec{\mathcal{L}}\cdot\vec{A}) = \vec{L}(\varepsilon,\vec{a}) , \tag{3.21}$$

since $\vec{\mathcal{L}}\,\varepsilon$ is zero from (3.13) and $\vec{\mathcal{L}}$ reduces to L on scalars. From (3.21) it follows that the second integral in (3.19) vanishes and hence

$$M = \int d\Omega(\varepsilon,\vec{n}\cdot\vec{\lambda}) \leq \int d\Omega|\vec{n}\cdot\vec{\lambda}| . \tag{3.22}$$

What is required therefore is a unit bound on $(\vec{n}\cdot\vec{\lambda},\ \vec{n}\cdot\vec{\lambda})$. To obtain it we return to (3.11). Taking the inner product of (3.11) with n_i, the L_i term drops out and we obtain

$$\Lambda_\Lambda \vec{F} = \vec{n}\times\vec{F} \quad \text{where } \Lambda = \vec{\Lambda}\cdot\vec{n} \tag{3.23}$$

Since the two cross-products commute we then have

$$\Lambda_\Lambda(\Lambda_\Lambda \vec{F}) = \vec{n}\times\{\vec{n}\times(\vec{n}\times\vec{F})\} \tag{3.24}$$

or

$$(\Lambda,\Lambda)(\vec{n}\times\vec{F})-\Lambda(\Lambda,\vec{n}\times\vec{F}) = \vec{n}\times\vec{F}. \tag{3.25}$$

But from (3.23) the second term on the left hand side of (3.25) vanishes. Hence

$$\{(\Lambda,\Lambda)-1\}(\vec{n}\times\vec{F}) = 0. \tag{3.26}$$

It is easy to see that $\vec{n}\times\vec{F}$ cannot be identically zero unless $\vec{F}$ is identically zero or singular. It follows that $(\underset{\sim}{\Lambda},\underset{\sim}{\Lambda})$ is unity for all r and hence for $r\to\infty$. Thus $(\vec{n}\cdot\vec{\lambda},\ \vec{n}\cdot\vec{\lambda})$ is unity. Hence for G=SU(2) and for spherical symmetry as defined in (3.9) we have $M\leq 4\pi/e$.

<u>Result 2.</u>

We next wish to show that for radial separation, the result $M \leq 4\pi/e$ holds for any Yang-Mills group G. By radial separation we mean[15] fields $\Phi(x)$ and $A(x)$ satisfying

$$\Phi(x) = \phi(\Omega)\left(\frac{H(r)}{r}\right), \qquad \vec{A}(x) = \vec{a}(\Omega)\left(\frac{1-K(r)}{r}\right) \tag{3.27}$$

in the Landau gauge $\vec{\nabla}\cdot\vec{A}=0$. Of course, we could also consider the first equation in (3.38) as defining a gauge, in which case $\vec{\nabla}\cdot\vec{A}=0$ would be an extra condition, but we prefer to keep the symmetry between Φ and $\vec{A}$ by regarding (3.38) as the imposed conditions. The method of procedure then is simply to insert (3.38) in the field equations (1.21) and the gauge condition $\vec{\nabla}\cdot\vec{A}=0$. From the gauge condition one obtains the two equations

$$\vec{r}\cdot\vec{a} = \vec{\partial}\cdot\vec{a} = 0 \tag{3.28}$$

and after a lengthy calculation one obtains from the

field equations five angular equations

$$\begin{aligned}
&(a)\quad \partial^2\vec{a} = -L(L+1)\vec{a} &&(d)\quad a_i^\alpha = \frac{1}{n}(\phi, t^\alpha \partial_i \phi)\\
&(b)\quad \partial^2\phi = -\ell(\ell+1)\phi &&(e)\quad d_i f = 0 \qquad\qquad (3.29)\\
&(c)\quad [a_i[a_i, a_s]] = -\mu^2 a_s &&
\end{aligned}$$

where L, ℓ, μ^2, n are constants, and $f(\Omega)$ is the radial scalar field defined in (3.6), together with two radial equations similar to (1.22) but slightly more complicated (all in addition to the FE equation (2.24)). The angular equations illustrate the remark of lecture 1 that the satisfaction of the indicial equations is not at all trivial, since to satisfy them we have to satisfy (3.29).

Before using equations (3.29) to obtain a bound for $eM/4\pi$ let us first briefly analyze them. Eqns. (a) and (b) state that $\vec{a}$ and ϕ must be definite spherical harmonics. Eqn. (c) states that (no matter what the Yang-Mills group G) the fields $a_i(\Omega)$ and $[a_i(\Omega), a_j(\Omega)]$ form an SU(2) subalgebra for each fixed Ω i.e. $a_i(\Omega) \subset SU(2)_\Omega \subseteq G$. Eqn. (d) can be combined with (3.2) so that it takes the form

$$M_{\alpha\beta}\, a_i^\beta(\Omega) = e^2 c^2\, n\, a_i^\alpha(\Omega), \tag{3.30}$$

which shows that the fields $a_i(\Omega)$ have definite masses. Eqn. (e), which will be the critical equation for the analysis, implies that the radial scalar field $f(\Omega)$ is spherically symmetric.

The full analysis of equations (3.29) has been carried out elsewhere[15] and is quite complicated, so we shall only sketch the general method of procedure here. First from the gauge conditions (3.28) and from (3.29a) one sees at once that $\vec{a}(\Omega)$ is a gradient field i.e.

$$\vec{a}_\alpha(\Omega) = -\frac{2}{L(L+1)}\vec{L}\, u_\alpha(\Omega) \quad \text{where } u_\alpha(\Omega) = -\frac{1}{2}\vec{L}\cdot\vec{a}_\alpha(\Omega). \tag{3.31}$$

By using this equation, (3.29e) and (3.29c) one finds after a certain amount of algebraic manipulation that

$$f_\alpha(\Omega) = \kappa u_\alpha(\Omega) \quad \text{where } \kappa = \frac{2}{3}\left(2-\frac{\mu^2}{L(L+1)}\right). \tag{3.32}$$

One can actually use (3.30) at this point to fix the value of κ, which turns out to be unity. However, without determining κ, we can substitute (3.32) and (3.31) back into (3.29e) to obtain

$$\partial_i f = \frac{2}{\kappa L(L+1)}\,[L_i f, f], \tag{3.33}$$

which is a self-coupling equation for f. Now the magnetic charge is given by (3.12) where $\varepsilon = f/|f|$. Hence (3.33) is clearly the key equation. To solve it we note from (3.31) and (3.29) that $f^\alpha(\Omega)$ is a spherical harmonic of order L and hence has the expansion

$$f^\alpha(\Omega) = \sum_m t^\alpha_m Y^L_m(\Omega), \quad t^{*\alpha}_m = (-1)^m t^\alpha_{-m}, \tag{3.34}$$

where $Y^L_m(\Omega)$ are the usual spherical harmonics. Using (3.34) in (3.33) one can show, using standard harmonic analyses and some algebraic manipulation, that the $f^\alpha(\Omega)$ must span a simple Lie subalgebra $\mathcal{J}$ of the Lie algebra of G.

Now, Eq. (3.34) shows that the spatial rotation group SO(3) acts on the linear space $\mathcal{J}$ by its (2L+1)-dimensional representation, and since (3.33) has the same form for all Ω we see that the structure constants of $\mathcal{J}$ are invariant under spatial rotations. Thus the rotation group is a group of automorphisms of $\mathcal{J}$. But for a semi-simple Lie group every continuous automorphism is inner. Therefore there exists an SO(3) subalgebra, $SO(3)_I$ of $\mathcal{J}$, which implements the infinitesimal spatial rotations and hence acts irreducibly on $\mathcal{J}$. But a non-trivial irreducible $SO(3)_I$ action on $\mathcal{J}$ is impossible unless $\mathcal{J} = SO(3)_I$. It follows that the subalgebra $\mathcal{J}$ spanned by $f(\Omega)$ is an SO(3) subalgebra. Since dim. $\mathcal{J} = 2L+1$ it also follows that L = 1.

If there are several non-conjugated SO(3) subalgebras of G the Higgs mechanism will in general choose among them. Equations (3.33), (3.32) and (3.31) then yield, up to an arbitrary orthogonal choice of basis in $SO(3)_I$

$$f^{\alpha}(\Omega)/|f(\Omega)| = n^{\alpha} \quad \text{(or constant)} \tag{3.35}$$

Since for $SO(3)_I$ the magnetic charge is given by (3.12) and $\varepsilon(\Omega)$ can be identified with $f(\Omega)/|f(\Omega)|$ we see that one again $\varepsilon(\Omega)$ is the unit (or the trivial) map from S_2 to S_2 and hence $M \leq 4\pi|e$.

We have thus shown that spherical symmetry for SU(2), and radial separation for any compact group, limits the value of the magnetic charge to $4\pi|e$. A more positive way to state this, perhaps, is to say that, under the conditions of spherical symmetry for SU(2) and radial separation for any compact group, the 't Hooft-Polyakov solution is unique. In any case we see that to obtain higher values of the magnetic charge, more complicated Ansätze needs to be used. The equations for <u>axially</u> symmetric solitons have already been derived and studied[(17)]. Alternatively in seeking higher values of the magnetic charge, one may abandon the search for explicit solutions and try only to establish, as in ref. (16), the <u>existence</u> of such solutions. That is, one may try to establish the existence of solutions to the Dirichlet problem defined by the boundary values $\phi(\Omega)$ and $\vec{a}(\Omega)$ corresponding to $eM|4\pi > 1$.

REFERENCES

(1) G. Derrick, J. Math. Phys. <u>5</u>, 1252 (1964).

(2) G. 't Hooft, Nucl. Phys. <u>B79</u> 276 (1974).
A. Polyakov, JETP Lett. <u>20</u> 194 (1974).

(3) A. Belavin et al. Phys. Letters <u>59B</u>, 85 (1975).
E. Witten, Harvard Preprint, HUTP-76/A-172.

(4) H. Nielsen, P. Oleson, Nuclear Phys. <u>B61</u>, 45 (1973).

(5) J. Bernstein, Rev. Mod. Phys. <u>46</u>, 7 (1974).

(6) M. Prasad, C. Sommerfield, Phys. Rev. Lett. <u>35</u> 760 (1975).

(7) B. Julia, A. Zee, Phys. Rev. <u>D11</u> 2227 (1975).

(8) P. Dirac, Proc. Roy. Soc. <u>A133</u>, 60, (1934), Phys. Rev. 74, 817 (1948).

(9) J. Arafune, P. Freund, C. Goebel, J. math. Phys. <u>16</u>, 433, (1975).

(10) M. Monastyrskii, A Perelmov, JETP Lett. <u>21</u> 43 (1975).

(11) N. Steenrod, Topology of Fibre Bundles (Princeton, 1951),
S. Hu, Homotopy Theory, Acad. Press (N.Y. 1959),
D. Husemoller, Fibre Bundles, Mc Graw-Hill (N.Y.1966)
R. Bott, Battelle Rencontres, Benjamin (N.Y. 1968).

(12) A. Schwarz, Nucl. Phys. B112 358 (1976)
(13) F. Englert, P. Windey, Phys. Rev. D 14, 2728 (1975).
(14) A. Guth, E. Weinberg, Phys. Rev. D 14, 1660 (1975), L. O'Raifeartaigh, Nuovo Cim. Letters 18 1548 (1977).
(15) L. Michel, L. O'Raifeartaigh, K.C. Wali, Syracuse Univ. Preprint COO-3533-83 SU-4210-83 (1976). Physics Letters (in press) E. Cremmer, F. Shaposnik, J. Scherk, Physics Letters 65B, 78 (1976).
(16) V. Fateev, Yu. Tyupkin, A. Schwartz, Teor. Mat. Fizika 26, 397 (1976).
(17) J. Madore, CNRS Marseille Preprint, GRG Journ. (in press)

SOME REFERENCES TO QUANTIZATION OF SOLITONS

(1) L. Faddeev, JETP Lett 21, 64, (1975), Physics Reports (in press).
(2) R. Dashen, B. Hasslacher, A. Neveu, Phys. Rev. D10, 4114, 4130, 4138 (1974), D4, 3424, (1975), D12, 2443, (1975).
(3) J.-L. Gervais, A. Jevicki, B. Sakita, Phys. Rev. D12, 1038, (1975).
(4) N. Christ, T.D. Lee, Phys. Rev. D12 1606 (1975).
(5) E. Tomboulis, Phys. Rev. D12 1678 (1975).
(6) P. Hasenfratz and G. 't Hooft, Harvard Preprint 1976
(7) R. Jackiw and C. Rebbi, MIT preprint 524
(8) P. Vinciarelli, Nucl. Phys. B89 463, 493 (1975).
(9) K. Cahill, Phys. Lett. 53B 174 (1974) 56B 275 (1975) 64B 283 (1976).
(10) W. Bardeen et al. Phys. Rev. D11 1094 (1975).

UNSOLVED PROBLEMS IN THE THEORY OF NON-LINEAR WAVE EQUATIONS

Michael Reed

Department of Mathematics

Duke University

The purpose of these first two lectures is to describe several unsolved problems in the general theory of non-linear wave equations. By "general", I mean techniques which are not too sensitive to the form of the equation. So I will not discuss the inverse scattering method or the recent work on the Korteweg-de Vries or Sine-Gordon equations which exploits the special properties of those equations. Today I will lecture on the local and global existence theory and tomorrow on the scattering theory. A note of warning is in order perhaps. The problems have varying degrees of difficulty. Some of them are difficult indeed and have resisted the efforts of many analysts. Others are reasonably straightforward problems on which it should be easy to make substantial and interesting progress. Still others are so vaguely stated that they could be easy or hard depending on how they are interpreted.

EXISTENCE THEORY

Let me begin by sketching the usual approach to the non-linear Klein-Gordon equation,

$$u_{tt} - \Delta u + m^2 u = f(u) \qquad x \in \mathbb{R}^n \tag{1}$$

$$u(x,o) = f_1(x)$$

$$u_t(x,o) = f_2(x)$$

We begin by reformulating the problem as an ordinary differential equation for a Hilbert space valued function. First we set $v = u_t$ and rewrite (1) as a first order system

$$\frac{d}{dt}\begin{pmatrix} u \\ v \end{pmatrix} = \begin{pmatrix} 0 & I \\ \Delta - m^2 & 0 \end{pmatrix}\begin{pmatrix} u \\ v \end{pmatrix} + \begin{pmatrix} 0 \\ f(u) \end{pmatrix} \tag{2}$$

$$\begin{pmatrix} u(x,0) \\ v(x,0) \end{pmatrix} = \begin{pmatrix} f_1(x) \\ f_2(x) \end{pmatrix}$$

Now, for each t, define $\phi(t) = \begin{pmatrix} u(x,t) \\ v(x,t) \end{pmatrix}$ and let

$$A = i\begin{pmatrix} 0 & I \\ \Delta - m^2 & 0 \end{pmatrix}$$

$$J(\phi) = \begin{pmatrix} 0 \\ f(u) \end{pmatrix}$$

$$\phi_0 = \begin{pmatrix} f_1(x) \\ f_2(x) \end{pmatrix}$$

Then we can rewrite (2) as

$$\phi'(t) = -iA\phi(t) + J(\phi(t)) \tag{3}$$

$$\phi(0) = \phi_0$$

Notice that A and J are operators which take pairs of functions of x into pairs of functions of x. Thus (3) is an ordinary differential equation in t for a function $\phi(t)$ which takes values in a space of pairs of functions of x. Of course, if we want a Hilbert space problem we must choose an appropriate norm on pairs of functions of x. The operator $-\Delta + m^2$ is a positive and self-adjoint on $L^2(\mathbb{R}^n)$ so its square root

$$B = \sqrt{-\Delta + m^2}$$

is well-defined by the functional calculus. Let D(B) denote the domain of B and denote the L^2 norm by $\|\ \|_2$. Then we define the norm of a pair $\langle u(x), v(x)\rangle$ by

$$\|\langle u,v\rangle\|_{(1)}^2 = \|Bu\|_2^2 + \|v\|_2^2$$

and take as our Hilbert space

$$\mathcal{H}_1 = \{\langle u(x), v(x)\rangle \mid \|\langle u,v\rangle\|_{(1)} < \infty\}$$

Why have we chosen this Hilbert space? The reason is that A is a self-adjoint operator on $\mathcal{H}_1$ (on an appropriate domain). Thus if J is zero we just have the Schrödinger equation

$$\phi'(t) = -iA\phi(t) \qquad (4)$$
$$\phi(o) = \phi_o$$

which is solved by $\phi(t) = e^{-iAt}\phi_o$. The plan is to try to solve (3) by treating (3) as a perturbation of (4). To do this we write the integral equation corresponding to (3):

$$\phi(t) = e^{-itA}\phi_o + \int_o^t e^{-iA(t-s)} J(\phi(s))ds \qquad (5)$$

For J nice enough we can solve this integral equation at least for t small. The proof, which uses the contraction mapping principle, is essentially the same as the proof for ordinary differential equations; there are just some technical complications due to the fact that A is an unbounded operator. But, what do I mean by "J nice enough". Well, it is sufficient for J to be a Lipschitz mapping of $\mathcal{H}_{(1)}$ to itself. In fact, if we want to differentiate (5) to recover (3) we need more than this. But, in particular, we need that J take $\mathcal{H}_{(1)}$ <u>into</u> itself. Let us see what this requires in the special case

$$f(u) = -u^p$$

where p is some positive integer. Thus, if $\phi = \langle u,v\rangle$, and $\phi \in \mathcal{H}_{(1)}$, then we need $||J(\phi)||_{(1)} < \infty$ when

$$||J(\phi)||^2_{(1)} = ||\binom{o}{-u^p}||^2_{(1)} = ||u^p||^2_2$$

If n, the dimension of the space variables, is one or two, then

$$||u^p||^2_2 = \int |u(x)|^{2p}dx \le C||Bu||^{2p}_2 \qquad (6)$$

so $||J(\phi)||_{(1)}$ is finite since $\phi \in \mathcal{H}_{(1)}$. (6) is an example of a Sobolev inequality, a generalization of the fundamental theorem of calculus. (6) says that the L^2 norm of powers of u can be estimated in terms of the L^2 of Bu where Bu is (roughly speaking) one derivative of u. These Sobolev inequalities are very sensitive to n. (6) is true for all p in one or two dimensions and it is true for p = 3, but <u>not</u> for p = 5 or higher in three dimensions. In the cases where it is true the single inequality (6) allows one to carry through the local (time) existence proof for (5) in the case $f(u) = -u^p$.

As with all non-linear equations solutions may exist only for a short time. To show that they exist for all t (that is to show that they are global) one needs to prove an apriori estimate which guarantees that the solution does not blow up in finite time. On our case here this means that we must show that $||\phi(t)||_{(1)}$ does not go to infinity in finite time. We can do this as follows. Suppose that $f(u) = -u^p$. Then by differentiating and using (1), we can see that the energy

$$E(t) = \int_{\mathbb{R}^n} (\nabla u(x,t))^2 + u_t(x,t)^2 + m^2 u(x,t)^2 + \frac{1}{p+1} u(x,t)^{p+1} dx$$

is conserved, that is $E(t) = E(o)$. Notice that if $\phi(t) = \binom{u(v,t)}{u_t(x,t)}$ is a (real-valued) solution of (3), then the first three terms in $E(t)$ are just $||\phi(t)||^2_{(1)}$. Thus, if p is odd, the last term is positive so we conclude that

$$||\phi(t)||^2_{(1)} \leq E(o) < \infty$$

and therefore solutions are global. If we had $f(u) = +u^p$ then this argument wouldn't work because there would be a minus sign in front of $1/p+1$.

We have just sketched the standard approach to the existence theory of non-linear waves equations which originated in [2], [5], [12]. In a phrase, the method is: Sobolev inequalities for local existence, then energy inequalities for global existence. But if the Sobolev inequalities don't work as in the case $n = 3$, $p = 5$? The first thing one naturally tries to do is change the space on which one works. This helps, locally. If we take as our norm

$$||\langle u,v\rangle||^2_{(k)} = ||B^k u||^2_2 + ||B^{k-1} v||^2_2$$

then A is again self-adjoint on the corresponding Hilbert space $\mathcal{H}_{(k)}$ and, for large enough k, J takes $\mathcal{H}_{(k)}$ into itself and the local existence argument goes through. But, now if we want global existence we must show that $||\phi(t)||_{(k)}$ does not go to infinity in finite time.

$||\phi(t)||_{(1)}$ remains bounded as before but no one has been able to combine this fact with the differential equation to show that $||\phi(t)||_k$ does not blow up in finite time.

A second approach [6], [11], [15] is to cut off the interaction $-u^p$ much as in field theory. One defines a sequence of interactions $f^{(\ell)}(u)$ so that the Sobolev estimates and the Energy inequality method both work on $\mathcal{H}_{(1)}$ for each ℓ. Then as ℓ goes to infinity, one extracts from the sequence $\{u^{(\ell)}\}$ of corresponding solutions a weakly convergent subsequence. In this way one gets a global solution for the case $f(u) = -u^5$ but it is a weak solution (in an appropriate distributional sense).

So, we know that strong solutions exist locally and weak solutions exist globally.

Problem 1 Prove that $u_{tt} - \Delta u + m^2 u = -u^5$ has strong global solutions in three dimensions.

Why is this such an important problem? It is because there are many other equations (for example the Navier-Stokes equations in three dimensions) where one would like global solutions but the method of Sobolev estimates fails. If one could handle $-u^5$, where the difficulty is isolated and the equation simple, by avoiding Sobolev estimates, then I'm confident that one could handle many of these other equations as well.

Now let us consider what happens if there is a conserved energy which is not positive. For example suppose $f(u) = +u^p$. Then the conserved energy is

$$E(t) = \frac{1}{2}\int_{\mathbb{R}^n} (\nabla u)^2 + u_t^2 + m^2 u^2 - \frac{1}{p+1} u^{p+1} dx$$

Then one can imagine that the first three terms blow up and the last term blows up but they cancel in such a way that $E(t)$ remains constant. In fact, one can prove that this is the case for appropriate initial data. One can see why this is so by looking at the differential equation (1). For $u > o$, the force u^p tries to make u still more positive and for $u < o$, u^p tries to make u more negative (p odd).

But the lack of a positive conserved energy does not necessarily mean that solutions will blow up in finite time. For example, suppose that the non-linearity f(u) is so weak that

$$||f(u)||_2 \leq C||u||_2$$

Then $||J(u)||_{(1)} \leq C||\phi||_{(1)}$ and for any solution of (5) we have

$$||\phi(t)||_{(1)} \leq ||\phi_o||_{(1)} + C\int_o^t ||\phi(s)||_{(1)} ds$$

Iteration of this inequality yields

$$||\phi(t)||_{(1)} \leq ||\phi_o||_{(1)} e^{ct}$$

so we have global existence.

Another interesting example where the conserved energy is not positive is the coupled Dirac and Klein-Gordon equations.

$$\begin{aligned} (-i\gamma^{\mu}\partial_{\mu} + M)\Psi &= g\, u\, \Psi \\ u_{tt} - \Delta u + m^2 u &= g\overline{\Psi}\gamma^{o}\Psi \end{aligned} \tag{7}$$

Because of special properties of the γ's there is a conserved energy

$$E(t) = \int_{\mathbb{R}^n} \overline{\Psi}\gamma_o(i\sum_{k=1}^{3}\gamma^k\partial_k + M)\Psi + \int_{\mathbb{R}^3} (\nabla u)^2 + m^2u^2 + u_t^2 - g\int_{\mathbb{R}^n} u\overline{\Psi}\gamma_o\Psi$$

which is not positive because of the first and third terms. Furthermore this interaction is not weak in the sense above. Nevertheless, in <u>one</u> dimension Chadam [3] has given a clever argument using the conservation of charge

$$Q(t) = \int_{\infty}^{\infty} |\Psi_1(x,t)|^2 dx + \int_{-\infty}^{\infty} |\Psi_2(x,t)|^2 dx$$

to show that solutions are global. The problem is open in three dimensions. So, the second problem is

<u>Problem 2</u> Prove global existence for certain classical wave equations which do not have a positive conserved energy, by exploiting the existence of other conserved quantities.

One has only to open the physics (and engineering) literature to see what a wide range of problems are covered under problem 2. There is a related problem which is so interesting that it deserves to be set out in its own right. Namely, global solutions may exist for some but not all initial data. A trivial example is that in the case (1) with $f(u) = u^p$ global solutions <u>will</u> exist if the initial data are small enough. A more interesting example is the proof by Chadam and Glassey [4] that for the equations (7) in three dimensions

$$\int_{\mathbb{R}^3} |\Psi_1 - \bar{\Psi}_4|^2 + |\Psi_2 + \bar{\Psi}_3|^2 dx$$

is a conserved quantity. Thus, if $\Psi_1 = \bar{\Psi}_4$ and $\Psi_3 = -\bar{\Psi}_3$ initially the same will be true for all times and it follows from this that

$$\bar{\Psi}\gamma_o\Psi = |\Psi_2|^2 + |\Psi_2|^2 - |\Psi_3|^2 - |\Psi_4|^2 = o$$

Thus the second equation becomes free. We can solve it and then complete the problem by solving the external field equation for the Dirac equation. This suggests:

<u>Problem 3</u> Consider (1) where $u = \langle u_1, \ldots, u_n \rangle$ is a vector of functions. Find out how the subspace of initial data for which global solution exist is related to the group of internal symmetries of f(u).

Because of the difficulties described above there has been relatively little work on properties of solutions in cases when there is global existence. For example is the solution an analytic function of various parameters of the theory (for example the coupling constant)? Can one exploit this to construct approximate computational schemes for various quantities of interest? Another example of "properties of solutions" is recent work by Velo and co-workers on dynamical charges and symmetry breaking for (1) in the case where the energy is only locally finite.

<u>Problem 4</u> Investigate properties of solutions in cases where global existence is known.

A further example of what I have in mind by problem 4 will be provided by the discussion of the propagation singularities in the third lecture.

<u>Problem 5</u> Investigate the problems of local and global existence for the initial value problem where u(x,t) is an operator-valued function.

At first glance this looks like the problem "solve quantum field theory" but it is really quite different. For in field theory one wants $u(x,t)$ to satisfy the Wightman axioms and these are very strong restrictions. For example, $u(x,t)$ must be an operator-valued <u>distribution</u> and thus all the usual singularities of field theory enter. Furthermore, the connection of the field theory problem to the initial value problem is not clear. One way to try to do the field theory problem is of course to solve (1) in the interaction picture with the free field and conjugate momentum as initial data. So, one demands less in that only special data are involved but more in that extra properties of the solution are required. By problem 5 I mean the pure initial value problem. Notice that global existence will be difficult even if the corresponding scalar equation has a positive conserved energy. This can already be seen if one lets $u(x,t)$ take values in the 2×2 matrices. There is some work in this direction by di Mottoni and Tesei.

SCATTERING THEORY

Let us begin by asking what one should mean by a scattering theory for an equation like

$$u_{tt} - \Delta u + m^2 u = f(u) \qquad (7)$$

where we assume that $f(o) = o$.
In the vaguest sense, a scattering theory means an understanding of the asymptotic behaviour of solutions of (7) as $t \to \pm\infty$. Usually one interprets this to mean showing that the solutions of (7) look more and more like the solutions of some simpler problem as $t \to \pm\infty$. If one can make explicit calculations on the simpler problem then one can derive an explicit asymptotic description of the more difficult problem. In the case of equation (7), the most natural candidate for the simple problem is the linear equation

$$u_{tt} - \Delta u + m^2 u = o \qquad (8)$$

The reason is that if the simple problem isn't linear we probably can't calculate very much about it anyway. Furthermore, nice solutions of (8) satisfy

$$\sup_x |u(x,t)| \sim t^{-n/2}$$

where n is the number of space dimensions. <u>If</u> a solution of (7) obeys a similar decay condition then it should look more and more like a solution of (8) because asymptotically f(u) will be smaller than the linear terms as long as the degree of f is greater than one. We will come back to this "if" in a moment.

In order to formulate the asymptotic relationship between the linear and non-linear problems, let us consider the problem on an abstract level. Let $\mathcal{H}$ be a Hilbert space with norm $\| \ \|$, A a self-adjoint operator on $\mathcal{H}$, and J a non-linear mapping of $\mathcal{H}$ into itself. Assume that J satisfies enough nice properties so that

$$\phi(t) = e^{-iAt}\phi_o + \int_o^t e^{-iA(t-s)} J(\phi(o))ds \qquad (9)$$

has global solutions; i.e. for each $\phi_o \in \mathcal{H}$, there is a continuous $\mathcal{H}$-valued function $\phi(t)$ which satisfies (9). Let $U_o(t) = e^{-iAt}$ and denote the mapping $\phi_o \to \phi(t)$ by U(t). That is,

$$U(t)\phi_o = \phi(t)$$

U(t) is a group of non-linear operators on $\mathcal{H}$. We would like to find a set of scattering states $\Sigma_{scat} \subseteq \mathcal{H}$ so that:

1(a) For each $\phi_- \in \Sigma_{scat}$, there is a $\phi_o \in \Sigma_{scat}$ so that

$$\|U(t)\phi_o - U_o(t)\phi_-\| \to o \quad \text{as } t \to -\infty$$

Define the wave operator W_+ by

$$W_+ : \phi_- \to \phi_o$$

1(b) For each $\phi_+ \in \Sigma_{scat}$, there is a $\phi_1 \in \Sigma_{scat}$ so that

$$\|U(t)\phi_1 - U_o(t)\phi_+\| \to o \quad \text{as } t \to +\infty \qquad (1o)$$

Define the wave operator W_- by

$$W_- : \phi_+ \to \phi_1$$

(2) Range W_- = Range W_+

Condition (2) is called asymptotic completeness. If it holds we can define the scattering operator by

$$S = (W_-)^{-1}W_+$$

That is

$$S : \phi_- \to \phi_+$$

(1a) and (1b) are usually not too hard. (1a) just says that given a solution $U_o(t)\phi_-$ of the linear problem, there exists a solution $\phi(t)$ of the non-linear problem which looks like it as $t \to -\infty$. A similar statement holds for (1b). But (2) is very hard, because it says that $\phi(t)$ (constructed to look like $U_o(t)\phi_-$ as $t \to -\infty$) should look free as $t \to +\infty$ also. That is, $\phi(o)$ should be the ϕ_1 corresponding to some ϕ_+. To see why this is hard, let's consider the example $f(u) = -u^3$, $J\langle u,v\rangle = \langle o,f(u)\rangle$ discussed above. Since $U_o(t)$ is unitary, (1o) is the same as

$$||U_o(-t)U(t)\phi(o) - \phi_+|| \to o \quad \text{as } t \to +\infty$$

So to prove the existence of ϕ_+ we need to show that $U_o(-t)U(t)\phi(o)$ is a cauchy sequence as $t \to +\infty$. From (9) we have

$$||U_o(t_1)U(t_1)\phi(o) - U_o(-t_2)U(t_2)\phi(o)||$$

$$\leq \int_{t_1}^{t_2} ||J(\phi(s))||ds$$

$$\leq \int_{t_1}^{t_2} ||u(x,s)^3||_2 ds$$

$$\leq \frac{1}{m}\sqrt{E(o)} \int_{t_1}^{t_2} (\sup_x|u(x,s)|)^2 ds$$

Thus in order to prove the existence of ϕ_+ we need an apriori estimate which guarantees that the solution of the non-linear problem decays fast enough so that the integral on the right side goes to zero as $t_1, t_2 \to +\infty$. This is the hard nugget of the non-linear scattering problem: it corresponds to the underlined "if" above. If one wants a complete scattering theory one must solve this problem by proving an apriori estimate. This is what was done by Morawetz and Strauss in their landmark paper [8] for a certain class of f's (including $f(u)=-u^3$)

in three dimensions.

Problem 6 Prove the existence and completeness of the wave operators for other equations than those handled by Morawetz and Strauss.

There are many cases where the existence of the wave operators is known (often by using the iteration techniques described below) so it is the apriori estimate that is lacking. To see why this is so hard notice that the equation may have a soliton solution. This would be a solution of the non-linear equation which doesn't decay at all. That doesn't mean that these won't be a scattering theory. It just means that the ranges of W_+ and W_- won't be the whole space. This is analogous to the situation in quantum mechanics when the range of the wave operators (in the 2-body case) is the part of the Hilbert space orthogonal to the bound states of the interacting system. The difficulty is that, unlike the quantum mechanical case, there is usually no nice way (like orthogonality) to separate those data which decay and those which have a soliton part. This is partly due to the fact that Σ_{scat} is typically a Banach space.

Problem 7 Investigate the ranges of the wave operators in a case where solitons are present.

In the case handled by Morawetz and Strauss the ranges of the wave operators are the whole space (that is, all solutions of the non-linear problem with sufficiently nice initial data decay). The solution of the whole problem in a case where solitons are present would be a great advance. But almost any partial information on range $W_\pm$ in such a case would be interesting.

It is clear that the above problems are hard. However, there are some cases where one can get information by iteration techniques. These were developed in Segal [14] and Chadam [2]. We will briefly describe a formulation due to Strauss [16]. The basic idea is that if the degree of the non-linearity is high and if we start at $t = -\infty$ with "small" data, then the non-linear term should remain very small and thus not affect the equation very much. We will state one theorem in the abstract setting which implements this idea.

Let $\mathcal{H}$ be a Hilbert space with norm $||\cdot||$, A a self-adjoint operator on $\mathcal{H}$, J a non-linear mapping from $\mathcal{H}$ to itself. Suppose that there are auxially norms $||\cdot||_a$

and $||\cdot||_b$ on $\mathcal{H}$ which satisfy:

(i) $||\phi||_a \leq C||\phi|| \qquad \phi \in \mathcal{H}$

(ii) $||e^{-itA}\phi||_a \leq C|t|^{-d}||\phi||_b \qquad |t| \geq 1$

(iii) There exist $p>o$, $\delta>o$, and $q\geq 1$ with $dq>1$ so that

$$||J(\phi_1) - J(\phi_2)|| \leq \beta(||\phi_1||_a + ||\phi_2||_a)^q||\phi_1 - \phi_2||$$

$$||J(\phi_1) - J(\phi_2)||_b \leq \beta\{(||\phi_1||_a+||\phi_2||_a)^{q-1}||\phi_1-\phi_2||_a + (||\phi_1||_a + ||\phi_2||_a)^q||\phi_1 - \phi_2||\}$$

for all ϕ_1,ϕ_2, with $||\phi_i|| \leq \delta$. In case $q = 1$ we assume that β can be chosen small if δ is small.

Notice that these hypotheses only require decay of the free solutions. (iii) are just fancy Lipschitz conditions where we have control over the constants in terms of the norm in which free solutions decay. Now we can define scattering states and the scattering norm. First, for a continuous $\mathcal{H}$-valued function $\Psi(t)$ on we define

$$|||\Psi(t)||| = \sup_t||\Psi(t)|| + \sup_t(1+|t|)^d||\Psi(t)||_a$$

Now, set

$$\Sigma_{scat} = \{\phi \in \mathcal{H} \mid |||e^{-itA}\phi||| < \infty\}$$

with the norm

$$||\phi||_{scat} = |||e^{-itA}\phi|||$$

Σ_{scat} is a Banach subspace of $\mathcal{H}$ with a stronger topology. The reason for choosing this strange definition for Σ_{scat} is that we want to find solutions of (9) which are close to free solutions. But we only expect this to be true for free solutions which decay. The norm $||\cdot||_{scat}$ just picks out those initial data for which the free solution decays at $\pm\infty$. Now, we can state:

<u>Theorem</u> Let $\mathcal{H}$, $||\ ||$, A, J, $||\cdot||_a$, $||\cdot||_b$ satisfy the above hypotheses. Then there is an $\eta > o$ so that for

every $\phi_- \in \Sigma_{scat}$ with $||\phi_-||_{scat} \leq \eta$, (9) has a global solution $\phi(t)$, with $|||\phi(t)||| \leq 2\eta$, which satifies:

(a) for each t, $\phi(t) \in \Sigma_{scat}$

(b) $||\phi(t) - e^{-itA}\phi_-|| \to o$ as $t \to -\infty$

(c) there is a $\phi_+ \in \Sigma_{scat}$ with $||\phi_+||_{scat} \leq 2\eta$ so that

$||\phi(t) - e^{-itA}\phi_+|| \to o$ as $t \to +\infty$

(d) the map $S : \phi_- \to \phi_+$ is one to one and continuous.

Notice that this theorem does not require any energy inequalities. All that is required is that the degree of the nonlinearity (related to q) be high enough and the decay of <u>free</u> solutions be fast enough so that $dq > 1$.

<u>Example</u>: Consider the non-linear Schrödinger equation in one dimension:

$$u_t - iu_{xx} + ku^p = o$$

$$u(x,o) = f(x)$$

The corresponding free equation is

$$u_t - iu_{xx} = o$$

$$u(x,o) = f(x)$$

with $A = -\dfrac{d^2}{dx^2}$. The solution of the free problem is

$$u(x,t) = (4\pi it)^{-1/2} \int e^{i(x-y)^2/4t} f(y)dy$$

so we immediately have the decay estimate

$$||u(x,t)||_\infty \leq ct^{-1/2}||f||_1$$

Therefore, we choose

$$||u|| = ||Bu||_2$$

$$||u||_a = ||u||_\infty$$

$$||u||_b = ||u||_1$$

You can now check that the hypotheses (i), (ii), and (iii) are satisfied with $d = \frac{1}{2}$, and $q = p - 2$. Thus, in order to apply the theorem we need $q > 2$, that is $p > 4$.

For more discussion, examples, and proofs see [1o] and [16].

Problem 8 Apply these "small data" techniques to new equations.

There are many equations in the applied mathematics and applied physics literature to which these ideas can be fruitfully applied. The applications are not trivial because one must prove a decay estimate for solutions of the free equation and one must choose the norms right. Both problems are usually harder than in the above example.

Next, we remark that in the case where one has global existence one can often use iteration techniques to prove the existence of $W_{\pm}$ even when the data is not small. One just iterates on an interval $(-\infty,T)$ when T is close to $-\infty$ and then continues the solution to $t = o$ by global existence.

Problem 9 Extend the iteration techniques to cases where $dq \leq 1$.

There are many problems which are not covered by the above theorem because the hypotheses on d and q are too strong. For example, one does not know the existence of the wave operators for (7) with $f(u) = -u^3$ in one dimension ($d = \frac{1}{2}$ and $q = 1$). To handle cases like this one needs subtler iteration techniques and more sensitive estimates than those used in the proof of the above theorem. This brings us to the last problem:

Problem 1o Investigate properties of $W_{\pm}$ and S in cases where they are known to exist.

Since $W_{\pm}$ and S are non-linear operators, finding their properties is not necessarily easy. Let me illustrate what sort of properties I have in mind. Suppose that (7) is viewed as a system for $u = \langle u_1,\ldots,u_k\rangle$. It would be interesting to investigate the invariance properties of $W_{\pm}$, S in terms of the group of internal symmetries of f. Also, it would be nice to know if $W_{\pm}$,S are analytic functions of various parameters in f(u), and if so in what regions.

Just as in quantum mechanics it would be nice to show that the small data scattering operator can be expressed as

$$S = I + T$$

where T is small in a suitable sense. Or, if f is replaced by λf, to show that S may be written

$$S = I + \lambda T_1 + \lambda^2 T_2 + \ldots$$

where the T_n are "simple" non-linear operators. This would permit the approximate calculation of S. Strauss and Morawetz prove that their S operator is Lorentz invariant and an analytic mapping on the scattering states [9].

REFERENCES

1. Browder, F., "On non-linear wave equations", Matn. Zeit. 80 (1962) 249-264.
2. Chadam, J., "Asymptotic for $u = m^2u + G(x,t,u,u_x,u_t)$, I,II", Ann. Scuola Norm. Sup., Pisa, 26 (1972), 33-65, 65-95.
3. Chadam, J., "Global solutions of the Cauchy problem for the (classical) coupled Maxwell-Dirac equations in one space dimension", J. Func. Anal. 13 (1973), 173-184.
4. Chadam, J. and R. Glassey, "On certain global solutions of the Cauchy problem for the (classical) coupled Klein-Gordon-Dirac equations in one and three space dimensions", Arch. Rat. Mech. Anal. 54 (1974) 223-237.
5. Jürgens, K., "Das Anfangswertproblem im Großen für eine Klasse nicht-linearer Wellengleichungen", Math. Zeit. 77 (1961), 295-3o8.
6. Lions, J., "Une remarke sur les problèmes d'évolution non linéaires dans les domaines non cylindriques", Rev. Roumaine Math. Pure Appl. 9, 11-18.
7. Morawetz, C., "Time decay for the non-linear Klein-Gordon equation", Proc. Roy. Soc. A 3o6 (1968), 291-296.
8. Morawetz, C. and W. Strauss, "Decay and scattering of solutions of a nonlinear relativistic wave equation", Comm. Pure Appl. Math. 25 (1972), 1-31.
9. Morawetz, C. and W. Strauss, "On a non-linear scattering operator", Comm. Pure Appl. Math. 26 (1973), 47-54.
1o. Reed, M., "Abstract non-linear wave equations", Springer Lect. in Math. 5o7 (1976).

11. Segal, I., "The global Cauchy problem for a relativistic scalar field with power interaction", Bull. Soc. Math. France 91 (1963), 129-135.
12. Segal, I., "Non-linear semi-groups", Ann. Math. 78 (1963), 339-364.
13. Segal, I., "Quantization and dispersion for nonlinear relativistic equations", Proc. Conference on Math. Theory Elem. Part., MIT Press, Cambridge, 1966, 79-108.
14. Segal, I., "Dispersion for nonlinear relativistic equations, II", Ann. Sci. Ecole Norm. Sup. (4) I (1968), 459-497.
15. Strauss, W., "On weak solutions of semi-linear hyperbolic equations", Anais Acad. Brazil, Ciências 42 (1970), p. 645-651.
16. Strauss, W., "Nonlinear scattering theory", in Scattering Theory in Mathematical Physics, ed. J.A. Lavita and J.-P. Marchand, Riedel Pub., Holland, 1974, p. 53-78.
17. Von Wahl, W., "Decay estimates for nonlinear wave equations", J. Func. Anal. 9 (1972), 490-495.

PROPAGATION OF SINGULARITIES

Michael Reed
Department of Mathematics
Duke University

In the last two lectures, I have given you, I hope, some feeling for the state of the art in the general theory of non-linear wave equations. In this lecture I want to describe some recent work [1] of a more special nature. The theory of propagation of singularities for linear wave equations has had a long development and is well understood. Let's consider the same problem for the equation

$$\begin{aligned} u_{tt} - u_{xx} &= f(x,u,u_x,u_t) \\ u(x,o) &= f_1(x) \qquad\qquad (1) \\ u_t(x,o) &= f_2(x) \end{aligned}$$

in one dimension. By propagation of singularities one means the relation between the smoothness properties of $u(x,t)$ in $\mathbb{R}^2$ and the smoothness properties of f_1 and f_2 on $\mathbb{R}^1$. The result is that, as in the linear case, singularities propagate along the boundary of the lightcone, but not into its interior. Of course, since f is non-linear one cannot deal with distributions as one can in the linear case, so we must say what we mean by "singularity". By a solution of (1) we will mean a once continuously differentiable solution of the corresponding integral equation. We say that $u(x,t)$ is "smooth" at $<x,t>$ if it is C^∞ in a neighborhood of $<x,t>$ and "singular" otherwise. We always assume that $f_1(x)$ is continuously differentiable and $f_2(x)$ is continuous. For simplicity we suppose that $\langle f_1, f_2 \rangle$ have compact support. We say that $\langle f_1, f_2 \rangle$ is "smooth" at $x \in \mathbb{R}$ if both f_1 and f_2 are C^∞ in a neighborhood of x and "singular" otherwise.

<u>Theorem</u> Suppose that f is C^∞ and satisfies $f(x,o,o,o)=0$. Suppose that the integral equation corresponding to (1) has global solutions for all pairs $\langle f_1, f_2 \rangle$ $C_o^{(1)} \times C_o^{(2)}$ and that the $C^{(1)}$ norm of the solution depends only on $||Df_1||_\infty$, $||f_1||_\infty$, and $||f_2||_\infty$. Let K be the set where $\langle f_1, f_2 \rangle$ is singular. Then $u(x,t)$ is C^∞ except on the union of the boundaries of light cones in $\mathbb{R}^2$ with base points in K. And, where it is C^∞, u satisfies (1).

The proof is too long to sketch here but let me explain the idea in a simple situation. Suppose that the singular set K is an interval $[-a,a]$ and let the regions be labeled as in Figure 1 when we assume that the support of $\langle f_1, f_2 \rangle$ is in $[-b,b]$.

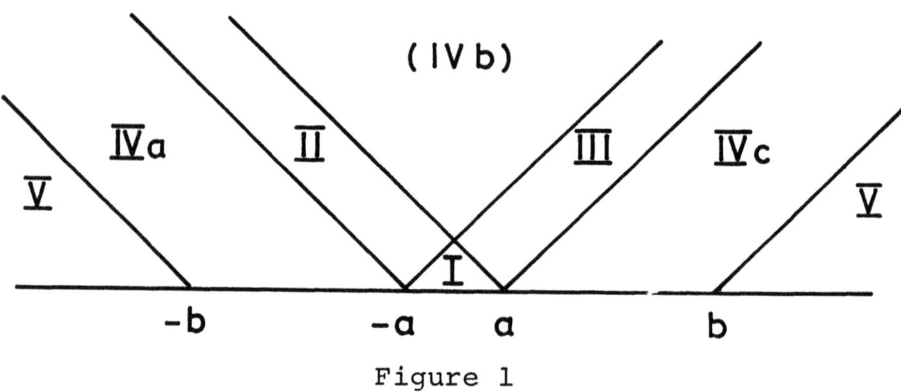

Figure 1

Then, since the solution propagates at speed one u will be zero in V and since the domain of dependence of a point in IVa or IVc does not include the singular interval K u should be C^∞ there. The point is to understand why u should be C^∞ in IVb. At any point $\langle x,t \rangle$ in IVb, $u(x,t)$ can be written as the free solution plus an integral over the boundary of the backward light cone from $\langle x,t \rangle$. The free solution is C^∞ in IVb. The integral involves u itself so the question is what differentiability properties should u have so that they are reproduced by the integral (so that one can apply the contraction mapping principle). The first guess is that u should be C^∞ in I $\cup$ II $\cup$ III but that is not right. The correct answer is the following. Let D_+ denote the directional denote the directional derivative in the direction $\hat{x} + \hat{t}$ where $\hat{x}$ and $\hat{t}$ are unit vectors in the x and t directions and let D_- be the directional derivative in the $-\hat{x} + \hat{t}$ direction. Then

(i) u will be C_1 in I

(ii) u will be C^∞ in the D_- direction in II but only C^1 in all other directions

(iii) u will be C^∞ in the D_+ direction in III but only C^1 in all other directions

To see why this is the right answer consider the one leg of the backward cone pictured in Figure 2.

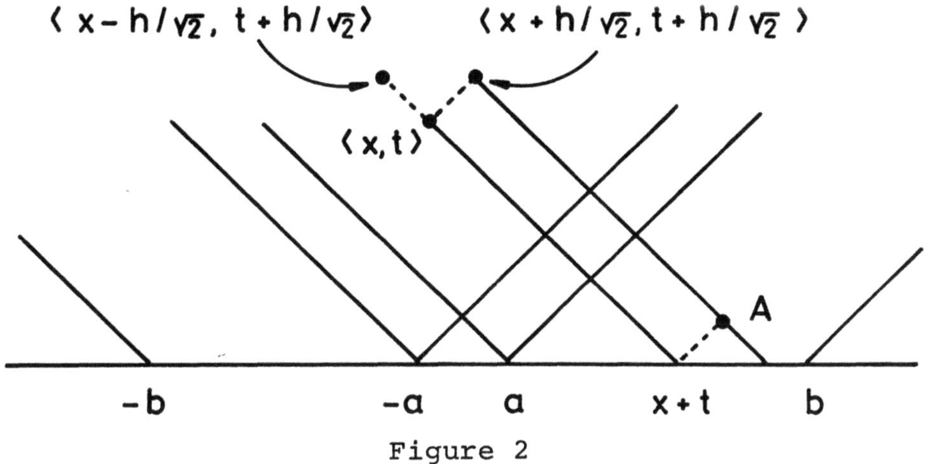

Figure 2

If one differentiates in the D_- direction, the difference quotient just gives the integral between $\langle x,t\rangle$ and $\langle x-h/\sqrt{2}, x+h/\sqrt{2}\rangle$ divided by h. The limit is just the integrand evaluated at $\langle x,t\rangle$ which is C^∞ there because u is. Thus the D_- direction is all right. The difference quotient in the D_+ direction is 1/h times the difference between the integral from $\langle x + t + 2h/\sqrt{2}, o\rangle$ to

$\langle x + h/\sqrt{2}, t + h/\sqrt{2}\rangle$ and the integral from $\langle x + t,o\rangle$ to $\langle x,t\rangle$. Dividing the first integral into two parts at A, one takes the limit and finds that $D_-u(x,t)$ is expressed as an integral along the original half cone of D_- applied to the integrand (which is all right because of (iii)) plus a contribution from the initial data at $\langle x + t,o\rangle$ which is also all right since the initial data is C^∞ there.

The details of the proof merely implement the above idea using standard iteration arguments and a domain of dependence estimate.

REFERENCE

[1] M. Reed, "Propagation of Singularities for Non-Linear Wave Equations in One Dimension", to appear.

DYNAMICS IN QUANTUM STATISTICAL MECHANICS

Derek W. Robinson

Université d'Aix-Marseille
II, Luminy Marseille
Centre de Physique Théorique, CNRS, Marseille

I. INTRODUCTION

In these lectures we describe and illustrate various general properties of dynamics in quantum statistical mechanics. Our principal interest is the construction of a dynamical description for a thermodynamic system. This construction corresponds essentially to integration of the equations of motion. In ordinary quantum mechanics the dynamics can be described by the Schrödinger equation determined by a symmetric operator H. The equation has a unique solution if, and only if, H is self-adjoint. Alternatively, one may work in the Heisenberg picture and the self-adjointness of H is necessary and sufficient to give a dynamical group acting on the algebra of observables. In statistical mechanics the Heisenberg picture appears to be more useful and one can often formulate the equations of motion in terms of a symmetric derivation δ acting on an algebra of observables. Integration of the equations of motion is then dependent on certain 'self-adjointness' properties of δ. Various possibilities occur depending on the system under consideration. The Heisenberg equations can lead to a dynamical group acting on a C^*-algebra of kinematic observables, and hence lead to a sensible evolution for every state of the system, or the equations might only yield solutions in situations close to equilibrium, and, of course, intermediate situations can occur.

For simplicity we mainly discuss these problems in the context of quantum spin systems but we attempt to present the material in a unified manner which is of use in more complex contexts.

We close with a few remarks concerning the properties of propagation.

2. LATTICE SYSTEMS

We first introduce the basic formulation of the lattice models which are to be subsequently analyzed. These models actually have two possible physical interpretations, either as lattice gases, or as spin systems. The first viewpoint interprets each point, $x \in \mathbb{Z}^\nu$, of the lattice as a possible site for a finite number N of particles, i.e. each point of the configuration space can be either empty or occupied by 1,2,...,N particles. These particles then interact with one another and this leads to an evolution in which the particles can be envisaged to jump from lattice site to lattice site. The second viewpoint is that every lattice site is permanently occupied by a particle but the particles have various internal degrees of freedom, e.g. the particles could have a spin with various possible orientations. The interaction between the particles then couples the internal degrees of freedom and this yields, for example, an evolution in which the spin orientations are constantly changing. The common quantum mechanical description of these models is given as follows:

First associate with each point $x \in \mathbb{Z}^\nu$ a Hilbert space $\mathcal{H}_x$ of finite dimension N+1 describing the states of the corresponding subsystem. If Λ is a finite subset of $\mathbb{Z}^\nu$ we associate with it the Hilbert space

$$\mathcal{H}_\Lambda = \prod_{x\in\Lambda}^{\otimes} \mathcal{H}_x$$

of dimension $(N+1)^{|\Lambda|}$ where $|\Lambda|$ is the number of points in Λ. We denote the algebra of matrices acting on $\mathcal{H}_\Lambda$ by $\mathfrak{A}_\Lambda$; $\mathfrak{A}_\Lambda$ is a finite-dimensional C* algebra. Note that if $\Lambda_1 \cap \Lambda_2 = \phi$

$$\mathcal{H}_{\Lambda_1 \cup \Lambda_2} = \mathcal{H}_{\Lambda_1} \otimes \mathcal{H}_{\Lambda_2}$$

We identify any operator $A \in \mathfrak{A}_{\Lambda_1}$ acting on $\mathcal{H}_{\Lambda_1}$ with the operator $A \otimes \mathbb{1}$ acting on $\mathcal{H}_{\Lambda_1 \cup \Lambda_2}$ where $\mathbb{1}$ denotes the identity operator acting on $\mathcal{H}_{\Lambda_2}$. This identification implies

$$\mathfrak{A}_{\Lambda_1} \subset \mathfrak{A}_{\Lambda_2}$$

whenever $\Lambda_1 \subset \Lambda_2$. As a result of this relationship the $\mathfrak{A}_\Lambda$ may be identified as subalgebras of an algebra $\hat{\mathfrak{A}}$ such that $\hat{\mathfrak{A}}$ is the union of all $\mathfrak{A}_\Lambda$. There is a natural norm on $\hat{\mathfrak{A}}$ introduced by the norms on the $\mathfrak{A}_\Lambda$ but $\hat{\mathfrak{A}}$ is not complete with respect to this norm. By completion we define a C*-algebra $\mathfrak{A}$ which represents the algebra of kinematic observables of our system, i.e. the $\mathfrak{A}_\Lambda$ correspond to the observables of the system Λ at a fixed time etc.

The algebra $\mathfrak{A}$ has a number of simple structural properties linked to its spatial structure. Let us introduce for each $a,x \in \mathbb{Z}^\nu$ a unitary mapping $V_x(a)$; $\mathcal{H}_x \to \mathcal{H}_{x+a}$ such that $V_x(o)$ is the identity on $\mathcal{H}_x$ and $V_x(a_1+a_2) = V_{x+a_2}(a_1)V_x(a_2)$. Define

$$V_\Lambda(a) = \prod_{x\in\Lambda}^{\otimes} V_x(a)$$

It is now possible to define an isomorphism α_x ; $\mathfrak{A}_\Lambda \to \mathfrak{A}_{\Lambda+x}$ for each $x \in \mathbb{Z}^\nu$ by

$$\alpha_x(A) = V_\Lambda(x)AV_{\Lambda+x}(-x)$$

This set of isomorphisms extends by continuity to a group $\mathbb{Z}^\nu$ of automorphisms of $\mathfrak{A}$.

The tensor product structure used in the construction of $\mathfrak{A}$ yields the commutation relations

$$[A,B] = o$$

for $A \in \mathfrak{A}_{\Lambda_1}$, $B \in \mathfrak{A}_{\Lambda_2}$, whenever $\Lambda_1 \cap \Lambda_2 = \emptyset$. It is then straightforward to conclude that

$$\lim_{|x|\to\infty} ||[A,\alpha_x(B)]|| = o$$

for all $A,B \in \mathfrak{A}$.

Physically the group α is interpreted as the group of space translations, i.e. translations in the configuration space, the commutation property expresses the independence of observables which are strictly localized in disjoint regions of space, and the asymptotic abelian property demonstrates that any pair of observations becomes independent when performed sufficiently far from one another.

Example 2.1 Assume N=1 then each $\mathcal{H}_x$ is two-dimensional and $\mathfrak{A}_x$ is an algebra of 2x2 matrices. It is well known that each such matrix can be decomposed as a sum of multiples of the identity matrix $\mathbb{1}$ and three Pauli matrices $\sigma_x = (\sigma_x^{(1)}, \sigma_x^{(2)}, \sigma_x^{(3)})$. The algebra of $2^{|\Lambda|} \times 2^{|\Lambda|}$ matrices $\mathfrak{A}$ then corresponds to the polynomials formed from the identity and the Pauli matrices $\{\sigma_x \,;\, x \in \Lambda\}$. The algebra $\mathfrak{A}$ is the set of all such polynomials together with some non-polynomial functions of the Pauli spin matrices. This model is often referred to as a spin - 1/2 system because in the spin interpretation each particle has only two degrees of spin freedom.

The foregoing structure defines the kinematics of our models. We next consider the dynamics. First we introduce the concepts of an interaction and a Hamiltonian. An interaction Φ is defined as a set of hermitian elements, $\Phi(X) \in \mathfrak{A}_X$, of $\mathfrak{A}$. We then associate with each finite subset $\Lambda \subset \mathbb{Z}^\nu$ a Hamiltonian $H_\Phi(\Lambda)$ by the definition

$$H_\Phi(\Lambda) = \sum_{X \subset \Lambda} \Phi(X)$$

The $\Phi(X)$ represents the interaction energy of particles localized at all points $x \in X$ and $H_\Phi(\Lambda)$ represents the total energy of interaction of particles contained in the subsystem Λ .

It is necessary to restrict the set of interactions Φ that we consider and a convenient method is to consider only those families B_ξ of Φ which satisfy

$$||\Phi||_\xi = \sup_{x \in \mathbb{Z}^\nu} \sum_{X \ni x} ||\Phi(X)||\xi(X) < +\infty$$

In this condition $X \subset \mathbb{Z}^\nu \to \xi(X) \in \mathbb{R}_+$ is a family of functions which will be further specified in applications. The $\xi(X)$ will always be taken as simple functions of the number of points $|X|$ in X and the diameter $D(X)$ of X,

$$D(X) = \sup_{x,y \in X} |x-y|$$

Mathematically these restrictions are very convenient because for each ξ the Φ, endowed with the linear structure inherited from $\mathfrak{A}$, i.e. $(\Phi_1+\Phi_2)(X) = \Phi_1(X) + \Phi_2(X)$, form a Banach space, $\mathcal{B}_\xi$, with respect to the norm $||\Phi||_\xi$. Physically such restrictions are not unnatural. For example if $\xi(X) = 1/|X|$ and we define

$$E_\Phi(x) = \sum_{X \ni x} \frac{1}{|X|} \Phi(X) \quad , \quad x \in \mathbb{Z}^\nu$$

then it follows that $E_\Phi(x) \in \mathfrak{A}$ and

$$||E_{\Phi}(x)|| \leq ||\Phi||_{\xi}$$

But $E_{\Phi}(x)$ corresponds to the energy of interaction of particles at the point $x \in \mathbb{Z}^{\nu}$ with their surroundings. Thus this choice of ξ corresponds to a restriction of finite energy per unit volume.

The Banach spaces $\mathcal{B}_{\xi}$ will always contain the set of all finite range interactions as a subspace. An interaction Φ is defined to be of finite range if there is a $d_{\Phi} > 0$ such that $\Phi(X) = 0$ whenever $D(X) > d_{\Phi}$. The minimum possible value of d_{Φ} for which this condition is satisfied is called the range of Φ. There is no mutual interaction between particles whose separation is greater than the range of Φ.

Example 2.2 Consider the spin - 1/2 system described in Example 2.1 and for simplicity assume $\nu=1$. Thus the system is formed by a one-dimensional chain of sites. The simplest form of interaction is given by assuming that o only $\Phi(\{x\}) \neq 0$ and $\Phi(\{x,x+1\}) \neq 0$. One such choice is

$$\Phi(\{x\}) = h\, \sigma_x^{(3)}$$

$$\Phi(\{x,x+1\}) = j_1\, \sigma_x^{(1)} \sigma_{x+1}^{(1)} + j_2\, \sigma_x^{(2)} \sigma_{x+1}^{(2)} + j_3\, \sigma_x^{(3)} \sigma_{x+1}^{(3)}$$

for some j_i, $h \in \mathbb{R}$. This model is usually referred to as the anisotropic Heisenberg model if the $j_i \neq 0$ and $j_i \neq j_j$ for some pair i,j; the isotropic Heisenberg model if $j_1=j_2=j_2\neq 0$; the X-Y model if $j_3=0$; the Ising model if $j_1=j_2=0$. In each case the energy of a system Λ consists of two parts

$$\sum_{x\in\Lambda} \sigma_x^{(2)} \quad , \quad \sum_{x,x+1\in\Lambda} \sum_{i=1}^{3} j_i\, \sigma_x^{(i)} \sigma_{x+1}^{(i)}$$

In the spin interpretation the first contribution corresponds to the interaction of the spin 1/2 particles with an external field and the second contribution corresponds to an energy of interaction of neighbouring spins.

Let us now consider the dynamics of the spin systems. The evolution with time of any system can be defined in terms of the evolution of the associated observables. For a finite lattice system confined to $\Lambda \subset \mathbb{Z}^{\nu}$, with interaction Φ, and Hamiltonian $H_{\Phi}(\Lambda)$, this evolution is given by

$$\tau_t^\Lambda (A) = e^{iH_\Phi(\Lambda)t} \ A \ e^{-iH_\Phi(\Lambda)t} \quad , \ t \in \mathbb{R}$$

for each observable $A \in \mathfrak{A}_\Lambda$. Thus $t \in \mathbb{R} \to \tau_t^\Lambda$ defines a one-parameter group of *-automorphisms of the matrix algebra $\mathfrak{A}_\Lambda$, e.g.

$$\tau_t^\Lambda(A^*) = \tau_t^\Lambda(A)^* \quad , \quad \tau_t^\Lambda(AB) = \tau_t^\Lambda(A) \quad \tau_t^\Lambda(B)$$

in correspondence with the Heisenberg picture of quantum mechanics. We now want to analyze this evolution for various classes of interactions and for systems which consist of a large number of particles. The standard method to approach this type of problem is by asymptotic expansion in terms of inverse powers of the total number of particles. The first term of such an expansion corresponds to the approximation that the system has an infinite number of particles.

Let us examine the nature of this approximation in more detail.

If we adopt the lattice gas interpretation of our model then each site is occupied by at most N particles. As there are $|\Lambda|$ sites the total number of particles N_Λ lies in the range $[o,, |\Lambda|N]$. In typical situations a gas will contain 10^{20} particles per unit volume and hence we are interested in very large values of N_Λ. To achieve such values we must allow $|\Lambda|$ to become very large and the asymptotic approximation $N_\Lambda = \infty$ then corresponds to a limit in which $|\Lambda| \to \infty$, i.e. the asymptotic approximation corresponds to the limit of an infinitely extended system. The computation of the time evolution of the lattice system then consists of calculating limits

$$\tau_t(A) = \lim_{\Lambda\to\infty} \tau_t^\Lambda(A)$$

of the evolutes $\tau_t^\Lambda(A)$ as the system grows indefinitely in size. We adopt for simplicity the convention that the limit $\Lambda\to\infty$ indicates that Λ eventually contains any finite subset of $\mathbb{Z}^\nu$. The initial question which must be answered concerning these limits is the sense in which they exist for a given interaction. This is one of the principal questions that we examine in the rest of these lectures. There are various possible senses in which the limits $\tau_t(A)$ might exist and the sense of the limit governs some of the basic properties of the evolutions $\tau_t(A)$, e.g. the continuity properties of the map $t\to\tau_t(A)$. The strongest sense that the limit might exist is in the topology defined by the norm on the algebra $\mathfrak{A}$ and we begin by examining this case.

3. C^*-ALGEBRA THEORY

Let the interaction Φ be fixed. In the previous sections we introduced the corresponding time evolution τ_t^Λ for a finite system Λ and it is now convenient to extend this definition from the $A \in \mathfrak{A}_\Lambda$ to all $A \in \mathfrak{A}$. Thus for any $A \in \mathfrak{A}$ we define $\tau_t^\Lambda(A)$ by

$$\tau_t^\Lambda(A) = e^{iH_\Lambda t} A e^{-iH_\Lambda t}$$

where we have now used the simplified notation $H_\Lambda = H_\Phi(\Lambda)$. Now we ask whether the $\tau_t^\Lambda(A)$ have limits, as $\Lambda \to \infty$, with respect to the norm on the C^*-algebra $\mathfrak{A}$, i.e. for each given choice of $A \in \mathfrak{A}$, $t \in \mathbb{R}$, and $\epsilon > o$ does there exist a $\Lambda(A,t,\epsilon) \subset \mathbb{Z}^\nu$ and a $\tau_t(A)$ such that

$$||\tau_t(A) - \tau_t^\Lambda(A)|| < \epsilon$$

for all $\Lambda > \Lambda(A,t,\epsilon)$. If this is indeed the case we call $\tau_t(A)$ the limit of $\tau_t^\Lambda(A)$ and write

$$\tau_t(A) = \lim_{\Lambda \to \infty} \tau_t^\Lambda(A)$$

Our immediate aim is to study such limits. Several simple remarks are in order.

Firstly note that the set of limits $\{\tau_t(A)\ ;\ A \in \mathfrak{A}\}$ is already determined by the set of limits $\{\tau_t(A); A \in \tilde{\mathfrak{A}}\}$ ($\tilde{\mathfrak{A}}$ is the algebra defined in section 2 as the union of the $\mathfrak{A}_\Lambda$). This follows because for each $A \in \mathfrak{A}$ and $\epsilon > o$ there is certainly a $B \in \tilde{\mathfrak{A}}$ such that

$$||A - B|| < \epsilon$$

But the definition of τ_t^Λ then ensures that

$$||\tau_t^\Lambda(A) - \tau_t^\Lambda(B)|| = ||\tau_t^\Lambda(A-B)|| < \epsilon$$

and hence the $\tau_t^\Lambda(A)$ form a Cauchy net with respect to the norm because

$$||\tau_t^{\Lambda_1}(A) - \tau_t^{\Lambda_2}(A)|| \le ||\tau_t^{\Lambda_1}(A) - \tau_t^{\Lambda_1}(B)||$$

$$+ ||\tau_t^{\Lambda_2}(A) - \tau_t^{\Lambda_2}(B)|| + ||\tau_t^{\Lambda_1}(B) - \tau_t^{\Lambda_2}(B)||$$

$$< 2\epsilon + ||\tau_t^{\Lambda_1}(B) - \tau_t^{\Lambda_2}(B)|| \ .$$

Thus it is only strictly necessary to examine the finite subsystems described by the $\mathfrak{A}_\Lambda$. The introduction of the global algebra $\mathfrak{A}$ is for technical convenience. Note, however, that even if $A \in \mathfrak{A}_\Lambda$ then its time evolute $\tau_t(A)$ is generally not an element of any $\mathfrak{A}_\Lambda$. Therefore the introduction of the algebra $\mathfrak{A}$ is very natural in the first asymptotic approximation of the time evolution.

Secondly we remark that if the $\tau_t(A)$ exist then they form a group of *automorphisms of $\mathfrak{A}$. One can check that for $\lambda,\mu \in \mathbb{C}$ and $A,B \in \mathfrak{A}$

$$\tau_t(\lambda A + \mu B) = \lambda\tau_t(A) + \mu\tau_t(B)$$

$$\tau_t(AB) = \tau_t(A)\ \tau_t(B)$$

$$\tau_t(A^*) = \tau_t(A)^*$$

$$\tau_s\tau_t(A) = \tau_{s+t}(A)$$

$$||\tau_t(A)|| = ||A|| \qquad \text{etc.}$$

The properties follow from similar properties of the $\tau_t^\Lambda(A)$. For example

$$||\tau_t(AB) - \tau_t(A)\ \tau_t(B)||$$

$$\leq ||\tau_t(AB) - \tau_t^\Lambda(AB)|| + ||\tau_t(A)\tau_t(B) - \tau_t^\Lambda(A)\tau_t^\Lambda(B)||$$

$$\leq ||\tau_t(AB) - \tau_t^\Lambda(AB)|| + ||(\tau_t(A) - \tau_t^\Lambda(A))\tau_t(B)||$$

$$+ ||\tau_t^\Lambda(A)(\tau_t(B) - \tau_t^\Lambda(B)||$$

$$\leq ||\tau_t(AB) - \tau_t^\Lambda(AB)|| + ||(\tau_t(A) - \tau_t^\Lambda(A))||\,||B||$$

$$+ ||A||\ ||\tau_t(B) - \tau_t^\Lambda(B)||$$

Taking the limit $\Lambda \to \infty$ the right hand side becomes zero by the definition of τ_t and hence we must have

$$\tau_t(AB) = \tau_t(A)\tau_t(B)$$

for each pair $A,B \in \mathfrak{A}$.

Thirdly we remark that the group τ_t of *automorphisms of $\mathfrak{A}$ is strongly continuous i.e. $t \in \mathbb{R} \to \tau_t$ is continuous in the sense that

$$\lim_{s \to o} ||\tau_{t+s}(A) - \tau_t(A)|| = o$$

This is again checked by a simple calculation which exploits the fact that the groups τ_t^Λ are strongly continuous and the continuity is preserved by the limiting process.

After these preliminary remarks we now attempt to establish circumstances in which the limits $\tau_t(A)$ actually exist. All the known methods of tackling this problem are either implicitly, or explicitly, dependent upon the equations of motion satisfied by the time evolutes. Before taking the limit $\Lambda \to \infty$ the evolutes $\tau_t(A)$ satisfy the Schrödinger type equation*

$$\frac{d\tau_t^\Lambda(A)}{dt} = \delta_\Lambda(\tau_t^\Lambda(A))$$

where the operator δ_Λ is defined for all $B \in \mathfrak{A}$ by

$$\delta_\Lambda(B) = i(-BH_\Lambda + H_\Lambda B) .$$

The basic method of examining the $\tau_t(A)$ in the limit $\Lambda \to \infty$ is to examine this Schrödinger equation in the limit. One tries to establish two things. Firstly one shows that the operators δ_Λ , acting on the C*algebra $\mathfrak{A}$, have, in some sense, an operator δ as a limit. Secondly one establishes that there exists a strongly continuous one-parameter group of *-automorphisms τ_t of $\mathfrak{A}$ satisfying the Schrödinger equation

$$\frac{d\tau_t(A)}{dt} = \delta(\tau_t(A))$$

If this is the case then it is quite natural to believe that τ_t will be the required limit of the τ_t^Λ and, under suitable circumstances, this can be established. In order to be more precise it is necessary to introduce some extra definitions.

*To have a direct analogy with the Schrodinger equation one should change the definition of δ_Λ by a factor -i. It is, however, notationally convenient to omit this factor.

A derivation of a C^*-algebra $\mathfrak{A}$ is defined to be a linear mapping from a dense *-subalgebra $D(\delta) \subseteq \mathfrak{A}$ the domain of δ, to a subspace $R(\delta) \subseteq \mathfrak{A}$, the range of δ, satisfying the property

$$\delta(AB) = \delta(A)B + A\delta(B)$$

for all pairs $A,B \in D(\delta)$. A derivation of this type is called symmetric if

$$\delta(A)^* = \delta(A^*)$$

for all $A \in D(\delta)$.

Example 3.1 Let Φ be an interaction of a quantum spin system and H_Λ the associated Hamiltonian for the finite subsystem Λ. Define δ_Λ by $D(\delta_\Lambda) = \mathfrak{A}$ and

$$\delta_\Lambda(A) = i\,(H_\Lambda A - A H_\Lambda)$$

then δ_Λ is a symmetric derivation of $\mathfrak{A}$. Note that the symmetry follows from our convention in placing the i. One has

$$\begin{aligned}\delta_\Lambda(A)^* &= -i\,(H_\Lambda A - A H_\Lambda)^* \\ &= i\,(H_\Lambda A^* - A^* H_\Lambda) = \delta_\Lambda(A^*)\end{aligned}$$

The principal interest of symmetric derivations of the above type is that they arise as infinitesimal generators of strongly continuous one-parameter groups of *-automorphisms of $\mathfrak{A}$. Let

$$A \in \mathfrak{A} \to \tau_t(A) \in \mathfrak{A}$$

be a one-parameter group of automorphisms of the foregoing type and define δ by

$$\delta(A) = \lim_{t \to 0} (\tau_t(A) - A)/t$$

where the limit is with respect to the norm on the algebra and $D(\delta)$ is the set of $A \in \mathfrak{A}$ such that the limit exists. It can be checked that δ is a derivation of $\mathfrak{A}$ satisfying the symmetry condition. We will refer to derivations which arise in the above manner as the infinitesimal generator of the group τ, or more briefly as the generator of τ; the automorphic and continuity properties of the group will be left implicit.

One basic property of infinitesimal generators is that they are densely defined. They are also closed in the Banach space sense, i.e. if $A_\alpha \in D(\delta)$ such that

$$\lim_\alpha ||A_\alpha - A|| = 0 \quad \text{and} \quad \lim_\alpha ||\delta(A_\alpha) - B|| = 0$$

then one automatically has $A \in D(\delta)$ and $\delta(A) = B$. We will give further characteristic properties below.

We are basically interested in properties of convergence of automorphism groups and this convergence can be characterized by convergence of certain functions of the generators. The most useful such function is the resolvent. In particular, if τ is a strongly continuous one-parameter group of *-automorphisms of the C*-algebra $\mathfrak{A}$, δ is the generator of τ, and $\alpha \in \mathbb{R}$ then one can define a linear operator

$$A \in \mathfrak{A} \to \frac{1}{1+\alpha\delta}(A) = \int_0^\infty dt\ e^{-t}\tau_{-\alpha t}(A)$$

(the integral is defined in the Riemann sense as a limit of sums approximating the integral in the norm sense). This definition is meaningful, because $||\tau_t(A)|| = ||A||$ etc., and the operator is denoted by $(1+\alpha\delta)^{-1}$ because one formally has $\tau_t = \exp\{t\delta\}$. The following criteria is often useful for convergence properties.

Proposition 3.1 (Kato-Trotter) Let $\{\tau_t^n\}$, and τ_t, denote strongly continuous one-parameter groups of *-automorphisms of a C*algebra $\mathfrak{A}$ and let $\{\delta_n\}$, and δ, denote the respective generators.

The following conditions are equivalent

1. $\lim_{n\to\infty} ||\tau_t^n(A) - \tau_t(A)|| = 0, \quad t \in \mathbb{R}, \ A \in \mathfrak{A}$

2. $\lim_{n\to\infty} ||(1+\alpha\delta_n)^{-1}(A) - (1+\alpha\delta)^{-1}(A)|| = 0,$ $\alpha \in \mathbb{R}, \ A \subset \mathfrak{A}$

Although this result provides a characterization of the type of convergence that we wish to study is often difficult to apply because the resolvent is not easily manageable; it replaces an unmanageable function $\exp\{t\delta\}$ with another $(1+\alpha\delta)^{-1}$. Thus it is often more useful to have a condition which is phrased directly in terms of the generators. The following is of use.

Proposition 3.2 Let $\{\delta_n\}$, and δ, be infinitesimal generators of strongly continuous one-parameter groups $\{\tau_t^n\}$, and τ_t, of *-automorphisms of a C*-algebra. Assume there exists a norm dense subspace D, of $\mathfrak{A}$, common to all the domains $\{D(\delta_n)\}$, and $D(\delta)$, of the generators. Furthermore assume that D is a core for δ.

It follows that if

$$\lim_{n\to\infty} ||\delta_n(A) - \delta(A)|| = o$$

for all $A \in D$ then

$$\lim_{n\to\infty} ||\tau_t^n(A) - \tau_t(A)|| = o$$

This result follows in a rather easy manner from the previous proposition. Recall that D is a core for δ means that if we restrict δ to the set D and then close the restricted operator then we reobtain δ, i.e. in symbols

$$\widetilde{\delta|_D} = \delta$$

Because δ is a generator one finds that D is a core if, and only if, the sets $r(1+\alpha\delta) = \{(1+\alpha\delta)(A);\ A \in D\}$ are dense in $\mathfrak{A}$ for each $\alpha \in \mathbb{R}$.

Proof Let $B \in r(1+\alpha\delta)$, i.e. $B=(1+\alpha\delta)(A)$ for some $A \in D$. One then has

$$\begin{aligned} &||(1+\alpha\delta_n)^{-1}(B) - (1+\alpha\delta)^{-1}(B)|| \\ &=||(1+\alpha\delta_n)^{-1}(1+\alpha\delta)(A) - A|| \\ &=||(1+\alpha\delta_n)^{-1}(\delta(A) - \delta_n(A))|| \\ &\leq||\delta(A) - \delta_n(A)|| \\ &\xrightarrow[n = \infty]{} 0 \end{aligned}$$

The last bound uses the estimate $||(1+\alpha\delta)^{-1}||\leq 1$ which is true for all generators δ. Thus one has convergence of the bounded resolvents $(1+\alpha\delta_n)^{-1}$, to $(1+\alpha\delta)^{-1}$, on a dense subset of $\mathfrak{A}$ and the general convergence follows immediately. Now Proposition 3.2 is a corollary of Proposition 3.1.

This last result, Proposition 3.2., is of frequent use in quantum statistical mechanics. For example consider a spin system with finite range interaction Φ. One can introduce a derivation δ of $\mathfrak{A}$ by $D(\delta) = \tilde{\mathfrak{A}}$ and

$$\lim_{\Lambda\to\infty} ||\delta(A) - \delta_\Lambda(A)|| = 0$$

for $A \in \tilde{\mathfrak{A}}$ because $\delta_\Lambda(A)$ is eventually independent of Λ. Now the convergence of the τ_t^Λ to a group τ_t is reduced to showing two properties of the operator δ obtained in this way.

1. δ can be closed
2. the closure $\bar{\delta}$, of δ , is a generator.

Thus we are forced to study criteria which determine whether a given derivation is closeable or whether it is a generator. A useful condition for closeability can be given in terms of invariant states.

Recall first that a state ω over a C^*-algebra is a linear functional which is positive, in the sense

$$\omega(A^*A) \geq 0$$

for all $A \in \mathfrak{A}$, and normalized to unity, i.e.

$$\sup_{A \in \mathfrak{A}} \omega(A^*A) \Big/ ||A||^2 = 1$$

(if the algebra contains an identity $\mathbb{1}$ then this latter condition is equivalent to $\omega(\mathbb{1}) = 1$).Moreover one can construct from each such state ω a representation of $\mathfrak{A}$ by bounded operators $\Pi_\omega(A),\ldots,$ on a Hilbert space $\mathcal{H}_\omega$ with a cyclic vector Ω_ω such that

$$\omega(A) = (\Omega_\omega, \Pi_\omega(A)\,\Omega_\omega)$$

The representation is called faithful if $\Pi_\omega(A) = 0$ implies that $A = 0$.

If τ_t is a group of automorphisms of $\mathfrak{A}$ with generator δ then the invariance of the state ω under τ, i.e. the condition

$$\omega(\tau_t(A)) = \omega(A) \qquad t \in \mathbb{R},\ A \in \mathfrak{A},$$

is equivalent to the infinitesimal invariance condition

$$\omega(\delta(A)) = 0 \qquad , A \in \mathfrak{A}.$$

One then has the following.

Theorem 3.1 (Bratteli-Robinson) Let δ be a symmetric derivation of a C*-algebra $\mathfrak{A}$. Assume there exists a state ω over $\mathfrak{A}$ which generates a faithful cyclic representation $(\mathcal{H}_\omega, \Pi_\omega, \Omega_\omega)$ and also satisfies the invariance condition

$$\omega(\delta(A)) = o \qquad , A \in D(\delta)$$

It follows that

1. δ is closeable
2. There exists a symmetric operator H on $\mathcal{H}$ such that

$$D(H) = \Pi_\omega(D(\delta))\Omega_\omega \quad \text{and}$$

$$\Pi_\omega(\delta(A))\psi = [iH, \Pi_\omega(A)]\psi$$

for all $A \in D(\delta)$ and $\psi \in D(H)$.

Now we turn to the characterization of generators.

Theorem 3.2 (Bratteli-Robinson) Let δ be a symmetric derivation of a C*-algebra $\mathfrak{A}$.

The following conditions are equivalent

1. δ is the infinitesimal generator of a strongly continuous one-parameter group of *automorphisms of
2. δ is closed,

   $$R(1+\alpha\delta) = \mathfrak{A}$$

   and $||(1+\alpha\delta)(A)|| \geq ||A||$, $A \in D(\delta)$
3. * δ is closed, δ possesses a dense set of analytic elements, and

---

* A is an analytic element for δ if $A \in D(\delta^n)$ for all $n = 1,2,\ldots$ and, moreover

$$\sum_{n\geq 1} \frac{|t|^n}{n!} ||\delta^n(A)|| < +\infty$$

for some $t \neq o$.

$$||(1+\alpha\delta)\ (A)|| \geq ||A|| \quad , A \in D(\delta)$$

Theorems 3.1 and 3.2 can be applied to the study of quantum spin systems and we consider several examples in the next section.

4. DYNAMICS OF QUANTUM SPIN SYSTEMS I

Throughout this section $\mathfrak{A}$ will denote the algebra of the quantum lattice gas and Φ an interaction of the class B_1, i.e. ξ is the unit function and

$$||\Phi||_1 = \sup_{x \in \mathbb{Z}^\nu} \sum_{X \ni x} ||\Phi(X)|| < +\infty$$

The Hamiltonians $H_\Lambda = H_\Phi(\Lambda)$ define derivations δ_Λ of $\mathfrak{A}$ by

$$\delta_\Lambda(A) = i\,[H_\Lambda\ A] \quad , A \in \mathfrak{A}$$

Moreover, we claim that there exists a derivation δ of $\mathfrak{A}$ such that

$$D(\delta) = \tilde{\mathfrak{A}} = \bigcup_\Lambda \mathfrak{A}_\Lambda$$

and

$$\lim_{\Lambda \to \infty} ||\delta(A) - \delta_\Lambda(A)|| = 0 \quad , A \in D(\delta)$$

To establish this one must first prove that the limit of the δ_Λ exists. But this is assured by the choice of $\Phi \in B_1$. For example if $A \in \mathfrak{A}_{\Lambda_o}$ one can identify δ by

$$\delta(A) = \sum_{X \cap \Lambda_o \neq \emptyset} [i\Phi(X), A]$$

and the infinite series is norm convergent because of the choice of Φ. Note that in the expressions $[H_\Lambda, A]$ the contributions $[\Phi(X), A]$ with $X \cap \Lambda_o = \emptyset$ are identically zero because of the commutation structure of $\mathfrak{A}$.

Our aim is to apply Proposition 3.2 to decide that the groups of automorphisms τ_t^Λ generated by the δ_Λ converge to the group τ. Thus we must show that δ is closeable and that its closure is a generator. First we consider closeability.

On the spin algebra $\mathfrak{A}$ there exists a special state, the trace state, ω_o which satisfies the property that

$$\omega_o(A) = \mathrm{Tr}_{\mathcal{H}_\Lambda}(A) \Big/ (N+1)^{|\Lambda|}$$

for all $A \in \mathfrak{A}_\Lambda$ and all $\Lambda \subset \mathbb{Z}^\nu$. The cyclicity of the trace then ensures that

$$\omega_o(\delta_\Lambda(A)) = o$$

for all $A \in \mathfrak{A}_{\Lambda_o}$ with $\Lambda_o \subseteq \Lambda$. Taking the limits $\Lambda \to \infty$ with Λ_o fixed one has

$$\omega_o(\delta(A)) = o$$

for all $A \in \mathfrak{A}_{\Lambda_o}$ and all $\Lambda_o \subset \mathbb{Z}^\nu$. Thus δ is closeable by Theorem 3.1.

Secondly consider the criteria that the closure $\tilde{\delta}$ of δ , is a generator. Both the criteria of Theorem 3.2 contain the resolvent bound condition

$$||(1+\alpha\delta)\ (A)|| \geq ||A|| \qquad , \alpha \in \mathbb{R}\ , \ A \in D(\delta)$$

(We call this condition a resolvent bound because it can be written as

$$||(1+\alpha\delta)^{-1}(A)|| \leq ||A||$$

for all $A \in D(1+\alpha\delta)^{-1}$). Let us check whether this condition is verified for $\tilde{\delta}$.

First remark that each δ_Λ is a generator because the associated groups τ_t^Λ are explicitly given by

$$\tau_t^\Lambda(A) = e^{iH_\Lambda t}\ A\ e^{-iH_\Lambda t} \quad , \quad A \in \mathfrak{A}$$

Thus the δ_Λ satisfy the resolvent bounds

$$||(1+\alpha\delta_\Lambda)(A)|| \geq ||A|| \qquad ,A \in \mathfrak{A}$$

Taking the limits one then has

$$||(1+\alpha\delta)(A)|| \geq ||A|| \qquad ,A \in D(\delta)$$

Finally this condition extends to the closure of δ by limiting and one has the desired conclusion

$$||(1+\alpha\tilde{\delta})(A)|| \geq ||A|| \qquad ,A \in D(\delta)$$

The information we have gathered so far can be summarized as follows.

Proposition 4.1 Let Φ be an interaction of a quantum spin system of the class B_1, i.e. such that

$$\sup_{x\in \mathbb{Z}^\nu} \sum_{X\ni x} ||\Phi(X)|| < +\infty$$

There exists a derivation δ of the quantum spin algebra $\mathfrak{A}$ such that

$$D(\delta) = \bigcup_{\Lambda\subset \mathbb{Z}^\nu} \mathfrak{A}_\Lambda$$

and

$$\delta(A) = i \sum_{X\cap\Lambda\neq o} [\Phi(X),A] \quad , \; A \in \mathfrak{A}_\Lambda$$

The derivation δ is closeable and its closure $\tilde{\delta}$ is the infinitesimal generator of a strongly continuous one-parameter group of *-automorphisms τ_t of $\mathfrak{A}$ if, and only if, one of the following two conditions is satisfied

<u>Either</u> $\mathcal{R}(1+\alpha\tilde{\delta}) = \mathfrak{A} \qquad \alpha \in \mathbb{R}$

<u>Or</u> $\tilde{\delta}$ possesses a dense set of analytic elements.

Finally, if $\tilde{\delta}$ generates the group τ then

$$\lim_{\Lambda\to\infty} ||\tau_t(A) - e^{iH_\Lambda t} A e^{-iH_\Lambda t}|| = o$$

for all $A \in \mathfrak{A}$.

Before continuing let us remark the following analogy. In usual quantum mechanics the dynamics are usually specified by a symmetric operator H acting on a Hilbert space $\mathcal{H}$. The dynamics are completely specified by a strongly continuous one-parameter group of unitary operators U_t on $\mathcal{H}$. The closure $i\hat{H}$, of iH, generates such a group if, and only if, $\hat{H}$ is self-adjoint, i.e. if, and only if, $i\hat{H}$ satisfies one of the conditions

<u>either</u> $\mathcal{R}(1+\alpha i\tilde{H}) = \mathcal{H} \qquad ,\alpha \in \mathbb{R}$

<u>or</u> $\tilde{H}$ possesses a dense set of analytic elements.

For quantum spin systems and interactions Φ of class B_1 the dynamics can be initially specified in terms of a symmetric derivation δ of the spin algebra $\mathfrak{A}$ and the dynamical description is complete if, and only if, the closure $\tilde{\delta}$ of δ satisfies the generalized criteria of self-adjointness

<u>either</u> $\mathcal{R}(1+\alpha\tilde{\delta}) = \mathfrak{A}$

<u>or</u> $\tilde{\delta}$ possesses a dense set of analytic elements.

Let us now give two examples which illustrate the applicability of these generalized self-adjointness criteria.

<u>Theorem 4.1</u> (Kishimoto) Let $\nu = 1$ and let Φ be an interaction of the class B_1. Further assume that

$$I_\Phi = \sup_{I} \sum_{\substack{X\cap I\neq\phi \\ X\cap I^c\neq\phi}} ||\Phi(X)|| < +\infty$$

where I denotes an interval of $\mathbb{Z}$ and I^c its complement.

It follows that the closure $\tilde{\delta}$ of the derivation δ defined by

$$D(\delta) = \bigcup_{\Lambda\subset\mathbb{Z}^\nu} \mathfrak{A}_\Lambda$$

and

$$\delta(A) = i \sum_{X\cap\Lambda\neq\phi} [\Phi(X),A] \qquad A \in \mathfrak{A}_\Lambda$$

satisfies the generalized self-adjointness criteria

$$R(1+\alpha\tilde{\delta}) = \mathfrak{A}$$

Hence $\tilde{\delta}$ generates a group τ and

$$\lim_{\Lambda\to\infty} ||\tau_t(A) - e^{iH_\Lambda t} Ae^{-iH_\Lambda t}|| = 0$$

for all $A \in \mathfrak{A}$.

Before giving the proof of the essential condition $\mathcal{R}(1+\alpha\tilde{\delta}) = \mathfrak{A}$ we remark that the extra condition imposed on the interaction assures that the energy of interaction between particles in I and particles outside of I is bounded uniformly for all I. This is not a reasonable type of condition to impose if $\nu\neq 1$ but it is easily satisfied if $\nu=1$. For example all finite range interactions satisfy the condition.

Proof Suppose that $\mathcal{R}(1+\alpha\tilde{\delta}) \neq \mathfrak{A}$. There must then exist a linear functional f such such $||f|| = 1$ and

$$f((1+\alpha\tilde{\delta})(A)) = 0$$

for all $A \in D(\tilde{\delta})$. In particular

$$f((1+\alpha\tilde{\delta})(A)) = 0$$

for all $A \in \mathfrak{A}_I$ and all intervals I. Now as $\mathfrak{A}_I$ is finite dimensional there are $A_I \in \mathfrak{A}_I$ with $||A_I||=1$ such that

$$f(A_I) = \sup_{\substack{A \in \mathfrak{A}_I \\ ||A||=1}} |f(A)| = ||f|\mathfrak{A}_I||$$

Therefore

$$f(e^{iH_I t} A_I e^{-iH_I t}) \leq f(A_I) = ||f|\mathfrak{A}_I||$$

and by the maximum principle

$$f(i[H_I, A_I]) = 0$$

Now denoting K_I by

$$K_I = \sum_{X \cap I \neq \phi} \Phi(X)$$

One has

$$\begin{aligned} 0 &= f((1+\alpha\tilde{\delta})(A_I)) \\ &= f(A_I) + \alpha f(i[K_I, A_I]) \\ &= ||f|\mathfrak{A}_I|| + \alpha f(i[K_I - H_I, A_I]) \end{aligned}$$

But

$$\begin{aligned} &|f(i[K_I - H_I, A_I])| \\ &\leq 2 \sup_I ||K_I - H_I|| \\ &= 2\, I_\Phi \end{aligned}$$

This gives a contradiction if $2|\alpha|I_\Phi < 1$. Hence $\mathcal{R}(1+\alpha\tilde{\delta}) = \mathfrak{A}$ for small $|\alpha|$. Thus within this range of α one has that $(1+\alpha\tilde{\delta})^{-1}$ is a bounded operator with $||(1+\alpha\delta)^{-1}|| \leq 1$.

Finally the range condition can be verified for all $\alpha \in \mathbb{R}$ by a standard Neumann series argument.

Theorem 4.2 Let Φ be an interaction of class B_ξ with $\xi(X) = \exp\{\lambda|X|\}$ for some $\lambda > o$.

Define a derivation δ by

$$D(\delta) = \bigcup_{\Lambda \subset \mathbb{Z}^\nu} \mathfrak{A}_\Lambda$$

and

$$\delta(A) = i \sum_{X \cap \Lambda \neq o} [\Phi(X), A] \quad , A \in \mathfrak{A}_\Lambda$$

It follows that δ has a dense set of analytic elements. In particular each $A \in \mathfrak{A}_\Lambda$ for some $\Lambda \subset \mathbb{R}^\nu$ is analytic for δ.

Therefore the closure $\tilde{\delta}$ of δ generates a group τ and

$$\lim_{\Lambda \to \infty} ||\tau_t(A) - e^{iH_\Lambda t} A e^{-iH_\Lambda t}|| = o$$

Proof Consider the perturbation series

$$\tau_t^{\Lambda,\Phi}(A) = A + \sum_{n \geq 1} \frac{(it)^n}{n} (\text{ad } H_\Lambda)^n A$$

where

$$(\text{ad}B)c = [B,c]$$

Using the definition of H_Φ and the local commutativity

$$(\text{ad}H_\Lambda)^n A = \sum_{X_1 \cap s_o = \phi \ldots . X_n \cap s_{n-1} = \phi} [\phi(X_n) \ldots [\phi(X_1), A]]$$

where

$$s_j = X_j \cup X_{j-1} \ldots . X_1 \cup \Lambda_o$$

Using

$$\sum_{X_i} = \sum_{n_i \geq 1} \sum_{X_i, |X_i| = n_i + 1}$$

and remarking that

$$|s_j| \leq |\Lambda_o| + n_1 + n_2 + \ldots + n_j$$

one finds straightforwardly that

$$||(\mathrm{ad}H_\Phi(\Lambda))^n A|| \le ||2^n \sum_{n_1,..n_n \ge 1} \prod_{i=1}^{n} (|\Lambda_o|+n_1+..+n_{i-1}) \, ||\Phi||_{n_i} \, ||A||$$

where

$$||\Phi||_{n_i} = \sup_{x \in \mathbb{Z}^\nu} \sum_{X_i \ni x, |X_i| = n_{i+1}} ||\Phi(X_i)||$$

But $x^n \le n!\lambda^{-n} e^{\lambda x}$ for $\lambda x \ge o$ and hence

$$||(\mathrm{ad}H_\Lambda)^n A|| \le ||A||(2/\lambda)^n ||\Phi||_\xi^n \, n! e^{\lambda|\Lambda_o|}$$

Thus the perturbation series converges uniformly for $2||\Phi||_\xi|t| < \lambda$ and this convergence is uniform in Λ. Hence $\bar{A}$ is analytic for δ.

5. GIBBS STATES

Although the theorems of the previous section establish the existence of an asymptotic dynamic description for many systems of interest, e.g. finite range interactions, they are unsatisfactory for several reasons.

There are interactions which have an important significance, at least theoretically, which are not covered by the theorems. For example in one-dimension, $\nu=1$, it is often the case that the description of equilibrium is unique, no phase transitions occur. The only known exceptions to this rule are systems with long range interactions. In particular, some spin systems with $\Phi \in B_1$ but such that $I_\Phi = \infty$. These models are not covered by Theorem 4.1.

A second drawback of the above results is that they are too strong, in a certain sense. The dynamics is determined as a group of automorphisms of the C* algebra of kinematic observables and its existence is independent of the state of the system. Each initial state of the asymptotic system would have a sensible, continuous, time evolution. But it is too much to hope that this will be the situation for all physical systems. It is easy to find continuous systems of particles which provide counter examples, e.g. the non-interacting Bose gas. There are particular pathological states which fail to have a continuous evolution. In the Bose gas these states assign a non-zero probability for finding an infinite number of particles in a finite region at some point in time. The moral of these examples is that it is impossible to hope that the dynamical evolution is always given by a strongly continuous group of *automorphisms of any kinematic C*algebra $\mathfrak{A}$. It is necessary to interpret the asymptotic equations of motion in some weaker sense.

There are various possibilities for constructing weak dynamics and there are many problems which remain to be solved and distinctions which remain to be made. We do not attempt to review all the known approaches but we describe some results which characterize the evolution of Gibbs equilibrium states.

For a spin system with interactions Φ confined to Λ the Gibbs equilibrium states are defined by

$$\omega_\Lambda(A) = \mathrm{Tr}_{\mathcal{H}_\Lambda}(e^{-\beta H_\Lambda}A) / \mathrm{Tr}_{\mathcal{H}_\Lambda}(e^{-\beta H_\Lambda})$$

where $A \in \mathfrak{A}_\Lambda$. Again one is interested in limits as $\Lambda\to\infty$ of these states and it is easily argued that certain limit points

$$\omega(A) = \widetilde{\lim_{\Lambda\to\infty}}\, \omega_\Lambda(A)$$

do exist for all $A \in \tilde{\mathfrak{A}}$. The ω then extend to states over $\mathfrak{A}$ by continuity. The set of such ω, plus the set of states obtained by modifying the H_Λ by surface terms, constitute the asymptotic approximations to the Gibbs equilibrium states. At some values of the inverse temperature β there will be several distinct limit points corresponding to several distinct thermodynamic phases. At other values of β there will be a unique limit point.

Associated to each limit point ω one has the canonical representation $(\mathcal{H}_\omega, \bar{\pi}_\omega, \Omega_\omega)$ and $\mathcal{H}_\omega$ is interpretable as the Hilbert space of states of the asymptotic system and Ω_ω the equilibrium vector. Of course the $\bar{\pi}_\omega(A), A\in\mathfrak{A}$, are still observables of the system.

The infinitesimal evolution of observables of the finite system is given by

$$A \to \delta_\Lambda(A) = [i\, H_\Lambda, A]$$

and this action can be reinterpreted in the representation $(\mathcal{H}_{\omega_\Lambda}, \Pi_{\omega_\Lambda}, \Omega_{\omega_\Lambda})$ associated with ω_Λ. One easily demonstrate that $\omega_\Lambda(\delta_\Lambda(A)) = 0$ for all $A \in \mathfrak{A}_\Lambda$ and hence there exists a bounded symmetric operator h_Λ on $\mathcal{H}_\Lambda$ such that

$$h_\Lambda\, \Omega_{\omega_\Lambda} = 0$$

and

$$ih_\Lambda \Pi_\omega(A)\Omega_{\omega_\Lambda} = \Pi_{\omega_\Lambda}(\delta_\Lambda(A)) \Omega_\omega$$

This is a direct consequence of Theorem 3.1. One also has

$$\omega_\Lambda(A \tau_t^\Lambda(B)) = (\Omega_{\omega_\Lambda}, \Pi_{\omega_\Lambda}(A) e^{ih_\Lambda t} \Pi_{\omega_\Lambda}(B)\Omega_\omega)$$

and h_Λ corresponds to the physical Hamiltonian of the finite system in the description yielded by $(\mathcal{H}_{\omega_\Lambda}, \Pi_{\omega_\Lambda}, \Omega_{\omega_\Lambda})$.

Now consider the limit point ω and assume that $\Phi \in B_1$. We have already remarked that the δ_Λ have a strong limit δ i.e.

$$\delta(A) = \lim_{\Lambda\to\infty} [i H_\Lambda, A] \quad , A \in \tilde{\mathfrak{A}}$$

and $\omega(\delta(A)) = o$ for all $A \in \tilde{\mathfrak{A}}$. Hence one can define a symmetric operator h on $\mathcal{H}_\omega$ by $D(h) = \{\Pi_\omega(A)\Omega_\omega , A \in \tilde{\mathfrak{A}}\}$ and

$$i\, h\, \Pi_\omega(A)\Omega_\omega = \Pi_\omega(\delta(A))\Omega_\omega$$

This is again an application of Theorem 3.1.

The operator h corresponds to a 'weak' limit of the Hamiltonians h_Λ e.g.

$$i(\Omega_\omega, \Pi_\omega(A)\, h\, \Pi_\omega(B)\Omega_\omega) = \tilde{\lim}\, \omega_\Lambda(A\delta_\Lambda(B)B)$$

$$= \tilde{\lim_{\Lambda\to\infty}} (\Omega_{\omega_\Lambda}, \Pi_{\omega_\Lambda}(A) h_\Lambda \Pi_{\omega_\Lambda}(B)\Omega_{\omega_\Lambda})$$

and it is natural to interpret h as the Hamiltonian of the asymptotic system. We emphasize that in general it has not been established that h has a self-adjoint closure and hence it is unclear whether it completely determines the evolution of the asymptotic, or thermodynamic system. But self-adjointness certainly suffices.

Theorem 5.1 (Bratteli-Robinson) Let Φ be an interaction of class B_1 and let ω denote a limit point of the Gibbs equilibrium states ω_Λ, let h denote the Hamiltonians introduced above within the representation $(\mathcal{H}_\omega, \Pi_\omega, \Omega_\omega)$. If h has a self-adjoint closure $\tilde{h}$ then

$$(\Omega_\omega, \Pi_\omega(A)\ e^{i\tilde{h}t}\Pi_\omega(B)\Omega_\omega)$$
$$= \widetilde{\lim_{\Lambda\to\infty}}\ \omega_\Lambda(A\tau_t^\Lambda(B))$$

Moreover

$$e^{i\tilde{h}t}\Pi''_\omega\ e^{-i\tilde{h}t} = \Pi''_\omega$$

where Π''_ω denotes the weak operator closure of the set Π_ω, i.e. h generates a weakly continuous one-parameter group of *automorphisms of the von Neumann algebra generated by $\Pi_\omega(\mathfrak{A})$.

This theorem shows that the self-adjointness of the thermodynamic Hamiltonian leads to an evolution of a similar nature, but of a weaker form, to that previously encountered. The Heisenberg evolution and the Heisenberg equations of motion are to be interpreted on the larger algebra Π''_ω. This von Neumann algebra can be viewed as the set of kinematic observables pertaining to the particular state ω of the system.

It is natural to conjecture that more is true than we have stated in the above theorem.

<u>Conjecture 5.2</u> Adopt the assumption of Theorem 5.1 including the fact that $\tilde{h}$ is self-adjoint.

It is conjectured that

$$\lim_{\Lambda\to\infty} ||(\Pi_\omega\ (\tau_t^\Lambda(A)) - e^{i\tilde{h}t}\ \Pi_\omega(A)e^{-i\tilde{h}t})\psi|| = o$$

for all $\psi \in \mathcal{H}_\omega$ and all $A \in \mathfrak{A}$ i.e. the $\tau_t^\Lambda(A)$ converge strongly in the representation.

In fact the strong convergence would be a consequence of weak convergence and the automorphism property because of the following simple result.

<u>Proposition 5.3</u> Let α_n be a sequence of *automorphisms of a von Neumann algebra $\mathcal{M}$ acting on the Hilbert space $\mathcal{H}$.

The following conditions are equivalent

1. The limits

$$\alpha(A) = \text{weak}\lim_{n\to\infty}\ \alpha_n(A)$$

exist for all $A \in \mathcal{M}$ and α is a *automorphism of $\mathcal{M}$.

2. The limits

$$\alpha(A) = \text{strong} \lim_{n\to\infty} \alpha_n(A)$$

exist for all $A \in \mathcal{M}$.

Thus what appears to be happening in the situation described by Theorem 5.1 is that the self-adjointness of $\tilde{h}$ ensures that the strong limits of the $\{\Pi_\omega(\tau_t^\Lambda(A))\,;A \in \mathfrak{A}\}$ exist as $\Lambda\to\infty$ and that these limits give automorphism groups of the von Neumann algebra Π_ω''.

Conversely the existence of the strong limits would be sufficient to ensure the automorphic property and the existence of a self-adjoint extension of h which determines the limit dynamics. Criteria for the existence of the strong limits can in fact be given as multiple limits

Theorem 5.4 Assume the following limits exist

$$\omega(A) = \lim_{\Lambda\to\infty} \omega_\Lambda(A)$$

$$G(A,B,C,D;t) = \lim_{\Lambda_1,\Lambda_2\to\infty} \lim_{\Lambda\to\infty} \omega_\Lambda(A\tau_t^{\Lambda_1}(B)\tau_t^{\Lambda_2}(C)\, D)$$

for all $A,B,C,D \in \mathfrak{A}$ and $t \in \mathbb{R}$.

The following conditions are equivalent

1. $\pi_\omega(\tau_t^\Lambda(A))$ converges strongly as $\Lambda \to \infty$.
2. $G(A,B,C,D;\, t) = \lim_{\Lambda_1\to\infty} \lim_{\Lambda\to\infty} \omega_\Lambda(A\tau_t^{\Lambda_1}(BC)D)$

for all $A,B,C,D \in \mathfrak{A}$ and $t \in \mathbb{R}$.

For this theorem to be true it is only necessary that the Gibbs state ω_Λ and the automorphisms τ^Λ are determined by the same Hamiltonian and it, of course, applies in a much wider context than that of quantum spin systems. Unfortunately, verification of condition 2 does not appear to be easy.

6. PROPAGATION PROPERTIES

In the previous sections we have shown how to construct, to first order, the time evolution of a large class of quantum spin systems. This is not, however, of great physical interest in itself. The physically interesting features are the properties of propagation, dispersion etc. of the evolution. Results in this direction are much sparser. We conclude these lectures by giving some fragmentary results.

Theorem 6.1 (Lieb-Robinson) Let Φ be an interaction of a quantum spin system of class B where $\xi(X)=|X|N^{2|X|}e^{\lambda D(X)}$ for some $\lambda > o$.

If $A,B \in \mathfrak{A}_{\{o\}}$ then

$$||[\tau_x\tau_t^{\Lambda}(A),B]|| \le 2||A||\,||B||\ \exp\{-\lambda|t|[|x|/|t|-||\Phi||_{\xi}/\lambda]\}$$

where τ_x denotes space-translations.

This result shows that the propagation due to τ_t^{Λ} or its limit τ_t is only non-negligible in the cone $|x|<|t|\ ||\Phi||_{\xi}/|\lambda|$. A more direct statement of the propagation into a cone is the following.

Theorem 6.2 Adopt the assumptions of Theorem 6.1 and define Λ_R by

$$\Lambda_R = \{x\ ;\ x \in \mathbb{Z}^{\nu}\ ,\ |x| \le R\}$$

For each $\epsilon > o$ and $V > 2||\Phi||_{\xi}/\lambda$ there is a D such that

$$\sup_{A \in \mathfrak{A}_{\{o\}}} ||\tau_t(A)-e^{iH_{\Phi}(\Lambda_{V|t|+D})}Ae^{-itH_{\Phi}(\Lambda_{V|t|+D})}||<\epsilon||A||$$

for all $t \in \mathbb{R}$. Thus $\tau_t(A)$, $A \in \mathfrak{A}_{\{o\}}$ is essentially localized in the conical region $\Lambda_{V|t|+D}$ for all $t \in \mathbb{R}$.

Although these results show that there is no essential propagation outside of a certain cone it is not automatic that there is any form of dispersion within the cone. A given class, B_{ξ}, of interactions contains both dispersive and non-dispersive interactions. One characterization of a dispersive interaction is the property of asymptotic abelianness, i.e.

$$\lim_{|t|\to\infty} ||[A,\tau_t(B)]|| = o$$

for all $A,B \in \mathfrak{A}$. Although this condition is both natural and useful (it has many consequences of an ergodic nature) it is apparently very difficult to verify for any given Φ. Thus it is logical to search for other characterizations. Various possibilities can be formulated in terms of the space-time algebras $\mathfrak{A}_{\Lambda,t}$ defined by

$$\mathfrak{A}_{\Lambda,t} = \{\tau_t(A) \; ; \; A \in \mathfrak{A}_\Lambda\}$$

We conclude with two examples.

Example 6.1 Consider the one-dimensional Heisenberg model of Example 2.2 with nearest neighbour interaction. Let $\mathfrak{A}^{\epsilon,x}$ denote the C^* algebra generated by the set of elements

$$\{\tau_t(A) \; ; \; t \in [o,\epsilon] \; , \; A \in \mathfrak{A}_{\{x,x+1\}}\}$$

It follows that

$$\mathfrak{A}^{\epsilon,x} = \mathfrak{A}$$

whenever two $j_i \neq o$ (Heisenberg or X-Y model) but

$$\mathfrak{A}^{\epsilon,x} \subset \mathfrak{A}_{\{x-1,x,x+1,x+2\}}$$

if only one $j_i \neq o$ (Ising model).

This property shows that the Ising model is definitely non-dispersive (elements $\tau_t(A)$, $A \in \mathfrak{A}_\Lambda$, are generally oscillatory) but indicates that there is a dispersion for the X-Y and Heisenberg models.

Example 6.2 Consider the X-Y model and define the algebra $\mathfrak{A}_{t,x}$ as the C*algebra generated by the set

$$\{\tau_{nt}(A) \; ; \; n \in \mathbb{Z}, \; A \in \mathfrak{A}_{\{x,x+1\}}\}$$

It follows that $\mathfrak{A}_{t,x} = \mathfrak{A}$ whenever $o < |t| < t_c$ but $\mathfrak{A}_{t,x} \neq \mathfrak{A}$ for $|t| > t_c$ where the critical value t_c is related to the group velocity V of the system $V = 1/t_c$.

Thus for one-dimensional models it appears natural to define Φ as dispersive if there is a finite interval $I \subset \mathbb{Z}^\nu$ and $t_c > o$ such that the C^*algebra generated by the set

$$\{\tau_t(A) \; ; \; t \in [o,\epsilon] \; , \; A \in \mathfrak{A}_I\}$$

is equal to $\mathfrak{A}$. Alternatively Φ is dispersive if there is an $I \in \mathbb{Z}^\nu$ and $t_c < o$ such that the C*algebra generated by the set

$$\{\tau_{nt}(A) \; ; \; n \in \mathbb{Z}, \; A \in \mathfrak{A}_I\}$$

is equal to $\mathfrak{A}$ for all $t < t_c$. In the latter case the group velocity should correspond to the inverse of the maximum possible t_c. It is unclear whether these two conditions of dispersiveness are equivalent and it is not known whether the Heisenberg model satisfies the latter condition.

BIBLIOGRAPHY

This short bibliography is intended as a guide to the literature which would aid further understanding of the foregoing lectures. We have purposefully selected only a small number of references and hope they will provide a starting point for broader reading on the subject. The general models of quantum spin systems that we have used as illustrations are described in further detail in

Ruelle, D.; Statistical Mechanics, Benjamin (New York)

Further details of the structure of their equilibrium states are described in

Lanford III,O.; Proceedings of the Cargese Summer School (1969) Gordon and Breach

More recent results are contained in

Araki, H.; Ion PDF; Commun.Math.Phys. 35,1 (1974)
Araki, H.; Commun.Math.Phys. 38,1 (1974)
Kishimoto, A.; Commun.Math.Phys. 47,167 (1976)

The general theory of derivations of C*algebras and the characterizations of generators has been developed by the author in collaboration with O.Bratteli (and partially with R.Herman)

Bratteli, O.; Robinson D.W.; Commun.Math.Phys. 42, 253 (1975)
Bratteli, O.; Robinson D.W.; Commun.Math.Phys. 46, 253 (1976)

The W*-(or von Neumann)algebra theory is described in

Bratteli, O.; Robinson,D.W.; Ann.Inst.Henri Poincaré Vol.XXV No.2 139 (1976)

The theory of unbounded derivations has been rigorously studied by numerous authers in the period 1974-1976 and references can be found by consulting the foregoing papers.

The Propositions 3.1 and 3.2 are simple generalizations of results for groups on Banach space. The theory of such groups is described in many places. For example

Kato, T.; Perturbation Theory, Springer Verlag, Berlin (1966)

The applications to spin systems Theorems 4.1 and 4.2 are extracted from

Kishimoto, A.; Commun.Math.Phys.47, 25 (1976)

Robinson, D.W.; Commun.Math.Phys. 7, 337 (1968)

Our discussion of the evolution of Gibbs states follows the more general discussion of

Bratteli, O.; Robinson, D.W.; Commun.Math.Phys.5o, 1331 (1976)

Earlier work on this time dependent Greens functions was done by M.B. Ruskai, M.Winnink, M. Sirugue, among others, and their work can be traced from the last reference. Criteria for strong convergence of the type occuring in Theorem 5.4 were originally given in

Dubin, D.A. and Sewell, G.; J.Math.Phys.II 299o (197o)

Theorem 6.1 was essentially proved in

Lieb, E. and Robinson, D.W.; Commun.Math.Phys. 28, 25 (1972)

This theorem was elaborated and applied in

Robinson, D.W.; Preprint Adelaide Univ.(1976)

The last paper which will appear in the Journal of the Australian Math.Soc., contains a discussion of various propagation properties.

SPONTANEOUSLY BROKEN SYMMETRY

R.F. Streater

Department of Mathematics

Bedford College

1. THE STANDARD MODEL

The wave-equation

$$\frac{\partial^2\phi}{\partial t^2} - \Delta\phi \equiv \quad \phi = o \tag{1.1}$$

is invariant under the transformations

$$\phi(\underline{x},t) \to \phi'(\underline{x},t) = \phi(\underline{x},t) + \eta, \quad \eta \in \mathbb{R} \tag{1.2}$$

That is, ϕ' obeys the same wave-equation, (1.1) as ϕ. Another way to say this is that the Lagrangean

$$\mathcal{L}(\phi(x),\chi(x)) = \tfrac{1}{2}\chi^\mu(x)\chi_\mu(x)+\chi^\mu(x)\partial_\mu\phi(x) \tag{1.3}$$

is invariant under

$$\begin{aligned} \phi &\to \phi' = \phi + \eta \\ \chi^\mu &\to \chi'^\mu = \chi^\mu \end{aligned} \tag{1.4}$$

i.e. $\mathcal{L}(\phi,\chi) = \mathcal{L}(\phi',\chi')$.
This is because $\mathcal{L}$ does not contain ϕ, only $\partial_\mu\phi$ i.e. ϕ is a cyclic coordinate. In such cases there is always a conserved current

$$\frac{\delta\mathcal{L}}{\delta(\partial_\mu\phi)} = \chi^\mu \tag{1.5}$$

and Lagranges' equations of motion are

$$\frac{\partial}{\partial x^\mu}\frac{\delta \mathcal{L}}{\delta(\partial_\mu\phi)} = \frac{\delta\mathcal{L}}{\delta\phi} \Rightarrow \partial_\mu\chi^\mu = o \qquad (1.6)$$

$$\frac{\partial}{\partial x^\mu}\frac{\delta \mathcal{L}}{\delta(\partial_\mu\chi_\nu)} = \frac{\delta\mathcal{L}}{\delta\chi_\mu} \Rightarrow o = \chi^\mu + \partial_\mu\phi \qquad (1.7)$$

Thus χ^μ is determined by a constraint equation, and combining (1.6) and (1.7) we get (1.1). The more usual $\mathcal{L} = \frac{1}{2}\partial_\mu\phi\partial^\mu\phi$ also leads to (1.1). On second quantization we may represent the field by an operator distribution $\hat{\phi}(x,t)$ on the usual Fock space, with vacuum $\underline{\Psi}_o$. Then

$$\begin{aligned} &\langle\underline{\Psi}_o, \hat{\phi}(x)\underline{\Psi}_o\rangle = o \\ &\langle\underline{\Psi}_o, \hat{\phi}'(x)\underline{\Psi}_o\rangle = \eta \end{aligned} \qquad (1.8)$$

We say that the symmetry (1.2) is <u>spontaneously broken</u>. More precisely, the fields $\hat{\phi}$ and $\hat{\phi}'$ are not unitary equivalent if $\eta \neq o$. This follows from (1.8). For suppose, if possible, that for some $\eta \neq o$ there is a unitary operator V such that

$$V\hat{\phi}(x)V^{-1} = \hat{\phi}'(x) = \hat{\phi}(x) + \eta \quad \text{for all } x \qquad (1.9)$$

Then, if $\underline{U}(t)$ denotes the unitary time-evolution on Fock space, $U_t\underline{\Psi}_o = \underline{\Psi}_o$ and

$$VU_{t'}\hat{\phi}(\underline{x},t)U_{t'}^{-1}V^{-1} = V\hat{\phi}(\underline{x},t+t')V^{-1} = \hat{\phi}(\underline{x},t+t') + \eta$$

$$= U_{t'}(\hat{\phi}(\underline{x},t)+\eta)U_{t'}^{-1} = U_{t'}V\hat{\phi}(\underline{x},t)V^{-1}U_{t'}^{-1}$$

Hence, $U_{t'}^{-1}V^{-1}U_{t'}V$ commutes with $\hat{\phi}(\underline{x},t)$ for all $\underline{x},t$. Since $\hat{\phi}$ is irreducible, Schur's lemma gives $U_{t'}^{-1}V^{-1}U_{t'}V = \lambda \mathbb{1}$ for some λ, for all t'. But then

$$U_{t'}^{-1}V^{-1}U_{t'}V\underline{\Psi}_o = \lambda\underline{\Psi}_o \qquad \text{giving}$$

$$U_{t'}(V\underline{\Psi}_o) = \lambda(V\underline{\Psi}_o)$$

Hence $V\underline{\Psi}_o$ is t-invariant (up to a factor). But $\underline{\Psi}_o$ is the only invariant vector under time-evolution, up to a factor. Hence $V\underline{\Psi}_o = \alpha\underline{\Psi}_o$ for some $\alpha \in \mathbb{C}$ (such that $|\alpha| = 1$). This leads to a contradiction

$$\eta = \langle \underline{\Psi}_o, \hat{\phi}'(\underline{x},t)\underline{\Psi}_o\rangle = \langle \underline{\Psi}_o, V\hat{\phi}(\underline{x},t)V^{-1}\underline{\Psi}_o\rangle =$$
$$|\alpha|^{-2}\langle \underline{\Psi}_o, \hat{\phi}\underline{\Psi}\rangle = o$$

This proves that no V exists.

The conserved current $\hat{\chi}_\mu = -\partial_\mu\hat{\phi}$ generates the symmetry in the sense that, if

$$\hat{Q} = \int d^3x\ \hat{\chi}^o(\underline{x},o) \quad , \quad V = e^{-iQ\eta} \tag{1.10}$$

then, by expanding V as a power series in η

$$V\hat{\phi}(\underline{x},t)V^{-1} = \hat{\phi}(\underline{x},t) + \eta \quad \text{for all } \underline{x},t \tag{1.11}$$

We now have a contradiction, since no such V exists. However, the integral in (1.10) does not converge in any sense strong enough to be able to conclude (1.11). Indeed, proceeding more carefully, we know that $\hat{\chi}^o(x) = \hat{\phi}(\underline{x},t)$ itself is not an operator; it is a distribution and needs 'smearing'; so that instead of $\hat{Q}$ in (1.1o) we may define, for each $\theta \in \mathcal{D}_r(\mathbb{R}^3)$, a self-adjoint operator

$$\hat{Q}(\theta) = \int \hat{\phi}(\underline{x},o)\,\theta(\underline{x})\,d^3x \tag{1.12}$$

and $$V(\theta) = e^{-i\eta\hat{Q}(\theta)} \tag{1.13}$$

Suppose $O_1 \subseteq \mathbb{R}^3$ is an open set on which $\theta(\underline{x}) = 1$, and $O_2 \supseteq O_1$ is the interior of the support of θ : $\bar{O}_2 = \{\underline{x} \in \mathbb{R}^3 : \theta(\underline{x}) \neq o\}$. Let $\hat{O}_1, \hat{O}_2$ denote the 'double cones' subtended by O_1 and O_2 in $\mathbb{R}^4$, located at $t = O$:

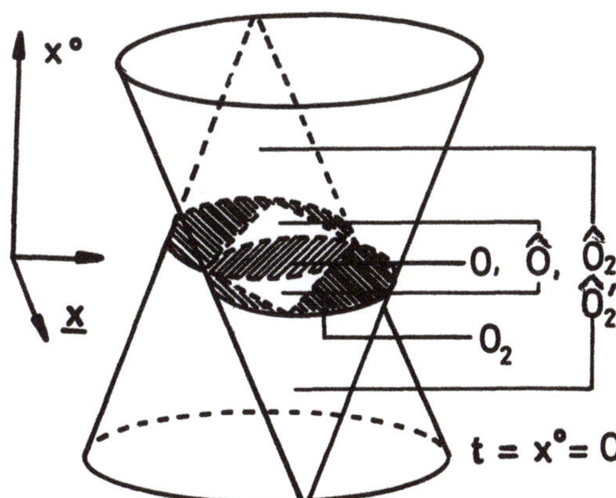

Figure 1. Ô is the <u>range of influence</u> of O propagating via the hyperbolic equation (1.1).

Then

$$V(\theta)\hat{\phi}(\underline{x},t)V^{-1}(\theta) = \hat{\phi}(\underline{x},t) + \eta, \quad x \in \hat{O}_1$$
$$= \hat{\phi}(\underline{x},t) \qquad x \in \hat{O}'_2 \tag{1.14}$$

where $\hat{O}'_2$ is the causal complement of O_2.
In the space between $\hat{O}_1$ and $\hat{O}_2^c = \mathbb{R}^4 - \hat{O}_2$, the transformation generated by V depends on the way θ behaves between O_1 and O_2. We see from (1.14) that while a V satisfying (1.9) does not exist, an operator doing the transformation for a limited range of $\underline{x},t$ does exist. We say the symmetry is <u>locally unitarily implemented</u>.

The infinitesimal form of (1.14) is

$$i[\hat{Q}(\theta),\hat{\phi}(\underline{x},t)] = 1 \qquad (\underline{x},t) \in O_1$$
$$= 0 \qquad (\underline{x},t) \in O'_2 \tag{1.15}$$

which follows from $\hat{\chi}^0 = -\hat{\dot{\phi}}$ and the canonical commutation relations. It follows from (1.14) and (1.15) that

$$\lim_{\theta\to 1} V(\theta)\hat{\phi}(\underline{x},t)V^{-1}(\theta) = \hat{\phi}'(\underline{x},t) \quad \text{for all } \underline{x},t$$
$$\lim_{\theta\to 1} i[\hat{Q}(\theta),\hat{\phi}(\underline{x},t)] = 1 \quad \text{for all } \underline{x},t \tag{1.16}$$

This is the sense in which (1.10) and (1.11) are to be understood.

In a Wightman theory in more than two dimensional space-time, we would not expect an interacting field, or current χ^μ, to "have a value" at sharp time. Thus, the expression

$$\left[\int d^3x\, \hat{\chi}^0(\underline{x},t)\theta(\underline{x}),\phi(\underline{y},t')\right] = A(t,t',\underline{y})$$

is an operator-valued distribution in $t,\underline{y},t'$, (rather than a function). But as such, it is independent of t if $(\underline{y},t') \in O_1$. To prove this, work formally:

$$\frac{\partial}{\partial t}A = \left[\int d^3x\, \frac{\partial\hat{\chi}^0}{\partial t}(\underline{x},t)\theta(\underline{x}),\phi(\underline{y},t')\right]$$
$$= [d^3x \text{ div } \vec{\chi}(\underline{x},t)\theta(\underline{x}),\phi(\underline{y},t')]$$

$$= -\left[\int d^3x\ \vec{\chi}(\underline{x},t)\nabla\theta(\underline{x}),\phi(\underline{y},t')\right]$$

Now, $\nabla\theta(\underline{x}) = 0$ on O_1, and by causality, the field $\vec{\chi}$ commutes with ϕ at space-like separation, i.e. if $(\underline{y},t')\in O_1$. Because of this in a general Wightman theory, if $\hat{\chi}^\mu$ is a conserved current, the operator

$$Q(\theta,\alpha) = \int \hat{\chi}^o(\underline{x},t)\,\theta(\underline{x})\,\alpha(t)\,d^4x$$

with $\int\alpha(t)dt = 1$, has the generator property

$$[Q(\theta,\alpha),\phi(\underline{y},t')] \quad \text{is independent of } \alpha,\ (\underline{y},t')\in\hat{O}_1$$

if supp α is small enough round O.

2. THE C^*-ALGEBRA FORMULATION

An alternative formulation of the above discussion is in terms of c^*-algebras. Let $\mathfrak{M}$ denote the set of real bounded c^∞ solutions to $\Box\phi = 0$; let $\mathfrak{M}_o$ denote those of which the Cauchy data $(\phi(\cdot,t),\dot{\phi}(\cdot,t))$ have compact support; let $\mathfrak{M}_1$ denote those for which the Poincaré invariant scalar product in ν space dimensions:

$$\langle\phi_1,\phi_2\rangle = \int\overline{\tilde{\phi}_1(\underline{p})}\,\tilde{\phi}_2(\underline{p})\,\frac{d^\nu p}{|\underline{p}|} + \int\overline{\tilde{\dot{\phi}}_1(\underline{p})}\,\tilde{\phi}_2(\underline{p})\,|\underline{p}|\,d^\nu p \qquad (2.1)$$
$$+\, i\int\{\dot{\phi}_1(\underline{x},o)\,\phi_2(\underline{x},o) - \dot{\phi}_2(\underline{x},o)\,\phi_1(\underline{x},o)\}d^\nu x$$

yields finite norm. Here, $\tilde{\phi}(\underline{p}) = \frac{1}{(2\pi)^{\nu/2}}\int e^{-i\underline{p}\cdot\underline{x}}\,\phi(\underline{x},o)d^\nu x$

$$\tilde{\dot{\phi}}(\underline{p}) = \frac{1}{(2\pi)^{\nu/2}}\int e^{-i\underline{p}\cdot\underline{x}}\,\dot{\phi}(\underline{x},o)d^\nu x.$$

In $\nu = 2$ or more space dimensions, $\mathfrak{M}_o \subseteq \mathfrak{M}_1$, but for $\nu = 1$, there are elements of $\mathfrak{M}_o$ not in $\mathfrak{M}_1$, e.g. if $\tilde{\dot{\phi}}(0) = q \neq 0$, then the first term in (2.1) diverges logarithmically. Note that, in our model, this occurs if

$$q = -\int\dot{\phi}(\underline{x},o)\,dx = \int\chi^o(x,o)\,dx \neq 0 \qquad (2.2)$$

i.e. if the total charge of the classical wave is non-zero. This fact, that $\mathfrak{M}_o \not\subseteq \mathfrak{M}_1$ if $\nu = 1$, is related to the fact that the free field of zero mass is not a Wightman field. This does not prevent us setting up a theory obeying Segal's and the Haag-Kastler axioms.

Let $\mathfrak{h}$ be the completion of $\mathfrak{M}_1$ in the norm, and furnish $\mathfrak{h}$ with a complex structure i.e. a multiplication by i, by

$$"i"(\phi(\underline{x},o),\dot{\phi}(\underline{x},o)) = (p\dot{\phi}, -\frac{1}{p}\phi) \tag{2.3}$$

Here, an element $\phi \in \mathfrak{M}$ is specified by its Cauchy data $(\phi(\underline{x},o),\dot{\phi}(\underline{x},o))$, and $p = \sqrt{-\Delta}$. Clearly, i maps $\mathfrak{M}_1$ to $\mathfrak{M}_1$ and $i^2 = -1$. One checks that $<,>$ is sesqui-linear in the complex sense. Indeed, one has:

Theorem $(\mathfrak{h}, i, <,>)$ is a complex Hilbert space, and the natural action of the Poincaré group, $\phi \xrightarrow{U_{a,\Lambda}} \phi_{a,\Lambda}$, where

$$\phi_{a,\Lambda}(x) = \phi(\Lambda^{-1}(x-a)) \tag{2.4}$$

provides a realization of the representation of mass 0 (and spin 0, if $\nu = 3$). That is, there is an isomorphism between $(\mathfrak{h}, i, <,>)$ and $L^2_{\mathbb{C}}(\mathbb{R}^\nu, \frac{d^\nu p}{|p|})$ that converts $U_{a,\Lambda}$ to the usual action of $\mathcal{P}$ on the positive light-cone. The proof of this theorem is a worthwhile exercise for anyone unfamiliar with this formalism.

The space $\mathfrak{h}$ is the natural test-function space in Segal's formalism. In the Haag-Kastler axiomatic scheme, we restrict ourselves to local operators which use the smaller space $\mathfrak{h} \cap \mathfrak{M}_o = \mathfrak{M}_o$ if $\nu > 1$. To each $\phi \in \mathfrak{h} \cap \mathfrak{M}_o$ define the self-adjoint operator on Fock space:

$$\{\hat{\phi},\phi\} = R(\phi) = \int \{\hat{\phi}(\underline{x},o)\dot{\phi}(\underline{x},o) - \dot{\hat{\phi}}(\underline{x},o)\phi(\underline{x},o)\}d^\nu x \tag{2.5}$$

This, the Wronskian between two solutions of the Klein-Gordon equation, is known to be Poincaré covariant, in that

$$\{\hat{\phi},\phi\} = \{\hat{\phi}_{a,\Lambda},\phi_{a,\Lambda}\} \tag{2.6}$$

Note that, in $R(\phi)$, $\hat{\phi}$ gets smeared with $\dot{\phi}$ and $\dot{\hat{\phi}}$ gets smeared with ϕ. Then $R(\phi)$ satisfies the Segal-Weyl relations: let $W(\phi) = e^{iR(\phi)}$. $W(\phi)$ is, for each ϕ, a unitary operator satisfying

$$W(\phi_1)W(\phi_2) = e^{i/2\ \mathrm{Im}<\phi_1,\phi_2>}\ W(\phi_1+\phi_2) \tag{2.7}$$

This method of quantization works for any linear canonical field. It fails for non-linear fields.

We may imagine the operator $W(\phi)$ located in any double cone $(\hat{O})$ in space-time based on a subset O of a space-like plane Σ such that the Cauchy data of ϕ on Σ are zero outside O. Each $W(\phi)$ is assigned in this way to many double cones, but never to two that are space-like separated. Let $\mathfrak{A}(\hat{O})$ denote the von Neumann algebra generated by all $W(\phi)$ with ϕ located in $\hat{O}$. Since products of $W(\phi)$'s are multiples of other $W(\phi)$'s by equation (2.7), the $*$-algebra generated by the $W(\phi)$'s is all operators of the form $\Sigma\alpha_i W(\phi_i)$. Then $\mathfrak{A}(\hat{O})$ is the closure of this in the strong operator limit, where ϕ_i's are located in $\hat{O}$. Denote by $U(a,\Lambda)$ the representation of $\mathcal{P}$ on Fock space. Then, because of the $\mathcal{P}$-covariance of the Wronskian, (2.6) one shows that, if $A \in \mathfrak{A}(O)$, then $U(a,\Lambda)AU^{-1}(a,\Lambda) \in \mathfrak{A}(O_{a,\Lambda})$ where $O_{a,\Lambda} = \{x : \Lambda^{-1}(x-a) \in O\}$ is the Lorentz transformed region. The system of algebras, $\hat{O} \rightarrow \mathfrak{A}(\hat{O})$, thus obey a version of the Haag-Kastler axioms.

Let $\mathfrak{A}$ be the c^*-algebra generated by all the $\mathfrak{A}(\hat{O})$, i.e. $\mathfrak{A}$ is the closure (in the sense of norm convergence) of $\bigcup_{O} \mathfrak{A}(\hat{O})$, ($\hat{O}$ bounded). An observable, or rather a local observable, is an element of $\mathfrak{A}$ that is self-adjoint. A <u>state</u> is a normalized positive linear functional on $\mathfrak{A}$. Each vector $\underline{\Psi}$ in Fock space defines a state $\rho_{\underline{\Psi}}$, namely, the positive linear functional

$$A \rightarrow \rho_{\underline{\Psi}}(A) = <\underline{\Psi},A\underline{\Psi}> \tag{2.8}$$

Such a state, $\rho_{\underline{\Psi}}$, is called a vector state. A density matrix ρ, i.e. a trace-class operator on Fock space of trace 1, defines a state, also called ρ:

$$\rho(A) = \mathrm{tr}(\rho A) \tag{2.9}$$

Such a state is impure (i.e. is a mixture) unless ρ is a 1-dimensional projection. A state defined by a density matrix is called a normal state. A pure normal state is a vector state. It is known that $\mathfrak{A}$ possesses states, even pures one, that are not vector states. Most of these extra states will not be of interest, but some are, for example, the vector states of theories with interaction. There may yet be some further states that can be interpreted as states of the free field, but states that carry a quantum number. The idea of Haag, Doplicher and Roberts is that these charged states should be obtainable from the Fock states by a <u>c^*-automorphism</u> of $\mathfrak{A}$.

A $*$-automorphism of a c^*-algebra $\mathfrak{A}$ is a 1 : 1 linear transformation $\tau : \mathfrak{A} \to \mathfrak{A}$ obeying

$$\tau(AB) = \tau(A)\tau(B) \tag{2.10}$$

$$\tau(A^*) = [\tau(A)]^* \tag{2.11}$$

For example, it is known that the Wronskian, $\mathrm{Im}\langle\phi_1,\phi_2\rangle$ between two solutions of the Klein-Gordon equation is invariant under a Poincaré transformation, $L \in \mathcal{P}$. Hence the commutation relations (2.7) are preserved by $\phi \to \phi_L$. One can then prove that there is a unique $*$ automorphism τ_L of the c^*-algebra generated by the $W(\phi)$ such that

$$\tau_L(W(\phi)) = W(\phi_L) \tag{2.12}$$

We say an automorphism τ is <u>spatial</u> if it is implemented by a unitary operator i.e. if there is a unitary V such that

$$\tau(A) = VAV^{-1} \quad \text{for all } A \in \mathfrak{A} \tag{2.13}$$

The operators $U(a,\Lambda)$ in Fock space implement $\tau_{a,\Lambda}$

$$U(a,\Lambda)W(\phi)U^{-1}(a,\Lambda) = W(\phi_{a,\Lambda}) \quad , \; L = (a,\Lambda) \tag{2.14}$$

Because of this, the automorphism τ_L can be extended to the strong closures of the algebras, i.e. to the von Neumann algebras $\mathfrak{A}(\hat{O})$, and so to $\mathfrak{A}$. Apart from the space-time symmetries defined by τ_L, a symmetry is, in this formalism, an automorphism that commutes with τ_t, the time-evolution.

Let us now translate our model of §1 into this formalism. The transformation (1.2) is $\hat{\phi}(\underline{x},t) \to \hat{\phi}(\underline{x},t)+\eta$ at any time. The algebra $\mathfrak{A}$ is generated by $\hat{\phi}$, $\hat{\dot{\phi}}$ at $t = o$. We may therefore specify the transformation of any element of $\mathfrak{A}$ by specifying the action on $\hat{\dot{\phi}}(f,o)$, $\hat{\phi}(g,o)$ to be

$$\begin{aligned} \hat{\phi}(f,o) &\to \hat{\phi}(f,o) + \eta\int f(x)d^{\nu}x \\ \hat{\dot{\phi}}(g,o) &\to \hat{\dot{\phi}}(g,o) \end{aligned} \qquad (2.13)$$

This determines the automorphism $W(\phi) \to W'(\phi) = \tau(W(\phi))$

$$W(\phi) \to e^{i\eta\int\dot{\phi}(\underline{x},o)d^{\nu}x}\, W(\phi) = e^{i\{\eta,\phi\}}\, W(\phi) \qquad (2.14)$$

where η is the (constant) solution to $\Box\,\eta = o$ having value η. Clearly, $W'(\phi)$ satisfies the Weyl relations (2.7), and so the map $A \to \tau(A)$ is an automorphism of the c^*-algebra generated by the $W(\phi)$, $\phi \in \mathfrak{M}_o$. For $\nu = 1$ we must choose $\phi \in \mathfrak{h} \wedge \mathfrak{M}_o$, and this requires $\int\phi d^{\nu}x = o$. Hence for $\nu = 1$ the automorphism reduces to the identity. We may define local charge operators by (1.12)

$$\hat{Q}(\theta) = \int \hat{\dot{\phi}}(\underline{x},o)\,\theta(\underline{x})\,d^{\nu}x$$

and then $V(\theta) = e^{-i\eta\hat{Q}(\theta)}$. Then, as in §1, we find that the automorphism is locally implemented by $V(\theta)$. In our case, a rigorous proof follows immediately from (2.7), once we realize that $V(\theta)$ itself is one of the $W(\phi)$'s. Because of local implementability, the automorphisms η extend to the weak closures $\mathfrak{A}(\hat{O})$ and hence to $\mathfrak{A}$.

The automorphism commutes with τ_L, and in particular with time-evolution. For

$$\eta \circ \tau_L W(\phi) = \eta W(\phi_L) = e^{i\{\eta,\phi_L\}} W(\phi_L)$$

$$= e^{i\{\eta_L{-1},\phi\}} W(\phi_L) = e^{i\{\eta,\phi\}} W(\phi_L)$$

$$= \tau_L\, e^{i\{\eta,\phi\}} W(\phi) = \tau_L \circ \eta\, [W(\phi)]$$

It is because of this that the formal field transformation $\hat{\phi} \to \hat{\phi} + \eta$ is consistently defined even though fields at later times are functions of $\hat{\phi}$, $\hat{\dot{\phi}}$ at $t = 0$ and the transformation at later times is already fixed by (3.13) at $t = o$.

We can prove that η is not spatial (implemented) (unless $\eta = o$), just as in §1. Thus, in c^*-language, a spontaneously broken symmetry is an automorphism, η, commuting with τ_t, but not implemented.

One often hears that, in a theory with a spontaneously broken symmetry, the vacuum is degenerate, i.e. multiple. This phenomenon is clearly understandable in c^*-language. If ρ is a state, and τ an automorphism of the c^*-algebra $\mathfrak{A}$, then $\rho \circ \tau$ is a new state i.e. $\rho \circ \tau(A) = \rho(\tau(A))$ defines a positive normalized linear functional on $\mathfrak{A}$. In this way, an automorphism acts (by duality) as a transformation on the set of states. The automorphism η of our model transforms the vacuum state, $\rho_{\underline{\Psi}_o}$, of Fock space, into a new state $\rho_{\underline{\Psi}_o} \circ \eta$. This is $\mathcal{P}$-invariant, and so can be called a vacuum state. But $\rho_{\underline{\Psi}_o} \neq \rho_{\underline{\Psi}_o} \circ \eta$ and so the theory has a continuum of vacua labelled by $\eta \in \mathbb{R}$.

Spontaneously broken symmetries can occur only in theories with ∞-many degrees of freedom, because only then do there exist automorphisms that are not spatial. This is closely related to the existence of many (indeed, ∞-many) inequivalent irreducible representations of $\mathfrak{A}$. A representation π, on a Hilbert space $\mathcal{H}$ is a linear map $\pi : \mathfrak{A} \to \mathcal{B}(\mathcal{H})$ (= set of bounded operators on $\mathcal{H}$) obeying

$$\begin{aligned} \pi(AB) &= \pi(A)\pi(B) \\ \pi(A)^* &= \pi(A^*) \end{aligned} \qquad (2.15)$$

Two representations are said to be equivalent if they are intertwined by a unitary operator. Thus if $(\pi_1, \mathcal{H}_1) \equiv (\pi_2, \mathcal{H}_2)$, then there exists a unitary operator $V: \mathcal{H}_1 \to \mathcal{H}_2$ such that

$$\pi_2(A) = V\pi_1(A)V^{-1}$$

Every state of $\mathfrak{A}$ is a vector state in some representation space (Gelfand-Naimark-Segal construction). Equivalent representations have the same vector states, and no state can be a vector state in two inequivalent representations. Thus the states of $\mathfrak{A}$ are divided up into classes, the vector states of the various different representations.

It is often said that two representations are inequivalent if their Hilbert spaces are orthogonal. This is not accurate since e.g. if $\pi = \pi_1 \oplus \pi_1$ acting on $\mathcal{H}_1 \oplus \mathcal{H}_1$, the vectors of the form $(\underline{\Psi}, o)$ and those of the form (o, Φ) are orthogonal, but the two subrepresentations are equivalent.

If $(\pi, \mathcal{H})$ is a representation, and τ an automorphism of $\mathfrak{A}$, then $\pi \circ \tau$, defined as operators on $\mathcal{H}$ by

$$[\pi \circ \tau](A) = \pi(\tau(A))$$

defines a new representation $(\pi \circ \tau, \mathcal{H})$ of $\mathfrak{A}$. $\pi \circ \tau$ is equivalent to π is and only if τ is spatial. In our example, π_F will denote the representation of $\mathfrak{A}$, on $\mathcal{H}$, by itself: $\pi_F(A) = A$. Clearly, (2.15) holds for π_F. The broken symmetry η leads to a continuum of representations

$$\pi_\eta(A) = \pi_F(\eta(A))$$

The vacuum $\rho_{\bar{\Psi}_0} \circ \eta$ is a vector state in π_η. It is immediately clear that τ_L, the Poincaré automorphisms, will be implemented in π_η. Indeed, since τ_L commutes with η, the $U(a,\Lambda)$ an $\mathcal{H}_F$ have the property

$$U(a,\Lambda)\pi_\eta(A)U^{-1}(a,\Lambda) = \pi_\eta(\tau_L(A))$$

for all $A \in \mathfrak{A}$, and $(a,\Lambda) \in \mathcal{P}$.
More generally, we can define a <u>covariant representation</u>, π as follows.

<u>Definition</u> If $\mathfrak{A}$ is a c^*-algebra and $L \to \tau_L$ a group of automorphisms of $\mathfrak{A}$ isomorphic to $\mathcal{P}$. Then a representation $(\pi, \mathcal{H})$ of $\mathfrak{A}$ is said to be <u>covariant</u> under $\mathcal{P}$ if there exists a continuous projective representation $U(L)$ of $\mathcal{P}$ on $\mathcal{H}$, such that

1) $U(L)\pi(A)U^{-1}(L) = \pi(\tau_L(A))$ for all $A \in \mathfrak{A}$ $L \in \mathcal{P}$

2) $U(a,1)$ has positive energy spectrum.

The multiple vacua need interpretation. In many body theory, where the phenomenon of spontaneous symmetry breakdown also occurs, the different vacuum states are physically different. However, in elementary particle physics, no two vacua are experimentally distinguishable. One can take the view that these theories of §1 must be dismissed as not agreeing with experiment. Or, one can reexamine measurement theory of the model by requiring that no physical difference should occur between $\rho_{\bar{\Psi}_0}$ and $\rho_{\bar{\Psi}_0} \circ \eta$. Thus we should take as our observables, not all of $\mathfrak{A}$, but the subset

$$\{A \in \mathfrak{A} : \rho_{\bar{\Psi}_0} \circ \eta(A) = \rho_{\bar{\Psi}_0}(A) \quad \text{for all } \eta\} \tag{2.16}$$

Sufficient for (2.16) is to take the set of elements of $\mathfrak{A}$ <u>invariant under η.</u> This is exactly the algebra generated by $-\partial_\mu\phi = \chi_\mu$ instead of ϕ. Thus, ϕ itself is a potential; the observable field is χ_μ, and $\phi \to \phi + \eta$ is a gauge transformation not touching the observables. The symmetry has disappeared altogether as a physical operation; instead, we have a spontaneously broken gauge symmetry. This is useful in constructing solitions.

3. GOLDSTONE'S THEOREM

Goldstone found that zero mass states are typical of a theory with a spontaneously broken symmetry. This was formulated into a general theorem by Goldstone, Salam and Weinberg who gave a non-rigorous but quite general proof in the relativistic setting.

In §1 the starting point of the broken symmetry is the existence of a current j^μ such that

$$Q(\theta \times \alpha) = \int j^o(\underline{x},t)\theta(\underline{x})\alpha(t)d^3xdt \tag{3.1}$$

where $\theta = 1$ on a large set, and $\int \alpha(t)dt = 1$, generates an infinitesimal transformation of $\hat{\phi} \to \hat{\phi} + \eta$, namely

$$i[Q(\theta \times \alpha), \hat{\phi}(y)] = 1 \qquad y \in \hat{O}_1 \tag{1.15}$$

More generally, this commutator will not be a c-number, e.g. if Q generates SO(2) rotations among fields ϕ_1, ϕ_2, then $[Q,\phi_1] = \phi_2$, and $[Q,\phi_2] = -\phi_1$. The symmetry is broken if ϕ_1 or ϕ_2 has non-zero expectation value. Even more generally, the object transformed into something with non-zero expectation value need not be linear in a Wightman field, but could be any local object. Thus, we take as our starting point for the Goldstone theorem the assumption that there exists a conserved current $j^\mu(\underline{x},t)$ and a local object A, commuting with $j^\mu(\underline{x},t)$ if $\underline{x}$ is large enough, such that, if $\alpha \in \mathcal{D}(\mathbb{R})$,

$$\langle \Psi_o, [\int j^o(\underline{x},t)\theta(\underline{x})\alpha(t)d^3xdt, A]\Psi_o\rangle = \eta \neq o \tag{3.2}$$

if $\int \alpha(t)dt = 1$ and $\theta(\underline{x}) = 1$ on a large enough set. This implies the presence of zero-mass particles.

<u>Proof</u> The distribution

$$\langle \Psi_o, [j^o(\underline{x},t), A]\Psi_o\rangle = F(\underline{x},t) \tag{3.3}$$

has compact support in $\underline{x}$ for fixed t, and its integral over space exists, $= \eta$, and is independent of time. Hence, its Fourier transform is, after smearing with α in p^o, an entire analytic function of $\underline{p}$. Its value at $\underline{p} = o$ is a distribution equal to $\eta\delta(p^o)$. Now, the spectrum of F is related to the spectrum of states in energy, because of the translation invariance of $j^o(\underline{x},t)$. Hence the spectrum of states (at $\underline{p} = o$) contains $\eta\delta(p^o)$ i.e. a mode of zero energy. This proof has been much debated. We must argue that the $\delta(p^o)$ term does not come from a vacuum state. Indeed, the vacuum contributions between the operators in the commutator cancel out i.e. if P_o is the projection, in the Hilbert space of the theory, onto the states of zero energy, then

$$\langle\underline{\psi}_o, AP_oB\underline{\psi}_o\rangle - \langle\underline{\psi}_o, BP_oA\underline{\psi}_o\rangle$$
$$= \langle\underline{\psi}_o, A\underline{\psi}_o\rangle\langle\underline{\psi}_o B\underline{\psi}_o\rangle - \langle\underline{\psi}_o, B\underline{\psi}_o\rangle\langle\underline{\psi}_o A\underline{\psi}_o\rangle = o$$

Clearly, the vacua in other representations cannot contribute. Thus it would seem that the term $\eta\delta(p^o)$ at $\underline{p} = o$ can only come from a contribution $\eta\delta(p^o - |\underline{p}|)$ at $\underline{p} = o$, i.e. from a zero mass state $\underline{\psi}$ such that $\langle\underline{\psi}_o, A\underline{\psi}\rangle \neq o$. A complete proof has been given by Ezawa and Swieca [4]. We shall give a short proof under the slightly different assumption that $A = \int\phi_\alpha(y)f(y)d^4y$, where ϕ_α is some Wightman field. The proof is then a rigorous version of that of Goldstone, Salam and Weinberg [3].

The distribution

$$\langle\underline{\psi}_o, [j^o(\underline{x},t), \phi_\alpha(\underline{y},t')]\underline{\psi}_o\rangle = F(x-y) \qquad (3.4)$$

is zero if x is space-like to $y = (\underline{y},t')$. Its Fourier transform $\tilde{F}(p)$ is zero unless $p^2 \geq o$, by the spectral condition. Then the remarkable theorem of Jost and Lehmann [5] ensures that F is a finite sum of derivatives of the usual commutator function (without assuming any specific Lorentz transformation laws for j^o, ϕ_α). Thus $\tilde{F}(p) = \sum p^{\mu_1}\ldots p^{\mu_n}\int\rho_n(k^2)\delta(p^2-k^2)\epsilon(p^o)dk^2$. At $\underline{p} = o$ we should get $\eta\delta(p^o)$. Thus,

$$\eta\delta(p^o) = \sum_{\text{some } j}(p^o)^j\int\rho_j(k^2)\delta(p^{o2}-k^2)\epsilon(p^o)dk^2$$

Only $j = 1$ can enter and all such terms can be combined to give

$$\eta\delta(p^o) = p^o\epsilon(p^o)\rho[(p^o)^2] = \eta|p^o|\delta[(p^o)^2]$$

Hence $\rho(k^2) = \eta\delta[(p^o)^2]$ and the theory has massless particles.

In constructive field theory we are just as likely to arrive at (3.4) as (3.3). Neither hypothesis is exactly an abstract statement about non-spatial automorphism groups - the group, if there is one, must be generated in the specific way implicit in the assumption that is generated by a Wightman conserved current. But the existence of such a current does not, by itself, ensure that an automorphism group can be defined - there are problems if $j^o(\theta \times \alpha)$ is not essentially self-adjoint or has not got enough analytic vectors.

Goldstone's theorem only works if a <u>continuous</u> symmetry is spontaneously broken. Glimm, Jaffe and Spencer have proved that, in strong coupling $(\phi)^4_{2,3}$ field theory, the symmetry $\phi \to -\phi$ is spontaneously broken, but the theory has a mass-gap.

The Goldstone theorem for relativistic theories prevents spontaneous breakdown of continuous symmetries from occurring in two dimensional space-time, at least in the form of (3.2) if $A = \phi(y)$. This remark was made by Coleman [6], but it immediately follows from the fact that the massless scalar field in two dimensions is not a Wightman field. For the free field, this is essentially the statement that $\mathfrak{M}_o \not\subset \mathfrak{H}$ (see §2). For a general scalar field, Wightman has emphasized this result [7], using that $\frac{dp}{|p|}$ is not a positive measure. Gal Ezer [8] notes that the same difficulty occurs if, for some A in (3.2) η is not zero. Thus, continuous symmetry groups cannot be broken if $\nu = 1$.

This does not mean that massless particles cannot exist in a two-dimensional Wightman theory. Indeed, as in §1, $\Box\phi = o$ defines a Wightman field $\chi^\mu = -\nabla_\mu\phi$.

A version of Goldstone's theorem can be proved in non-relativistic quantum field theory and in statistical mechanics. Instead of causality, (i.e. the exactly vanishing commutator at space-like separation) we might expect, for sufficiently short range forces, for the commutator, $[j^o(\underline{x},t),\phi(\underline{y},o)]$ to fall off sufficiently rapidly as $|\underline{x}-\underline{y}| \to \infty$, for each t, for us to be able to throw away the space-like surface integral at ∞ that arises from using $\partial_\mu j^\mu = o$. Unless we can, $[Q(\theta,t),\phi(y)]$ is not independent of t however large the region where $\theta = 1$. It is easy enough to postulate rapid fall-off for $[j^o(\underline{x},o),\phi(\underline{y},o)]$, but whether this fall-off is maintained for all time differences is a dynamical property which is hard to prove in specific models. The first serious treatment of a model is in [9] where the Heisenberg ferromagnet with short range forces is treated. Of course, one does not arrive at massless relativistic particles; one merely gets that there is no energy gap. One can prove this result by other means, and the theorem of Goldstone is less useful in non-relativistic theory than in Wightman theory. However, it remains a heuristic tool, and lends support to the curious connections between non-relativistic statistical mechanics and the Coleman-Gal-Ezer remark. Thus

1. Mermin shows that in two dimensional isotropic Heisenberg model, rotation symmetry is not spontaneously broken at non-zero temperatures

2. Dyson-Simon-Lieb show that it is spontaneously broken in 3 dimensions.

3. The discrete symmetry $\phi \to -\phi$ is spontaneously broken in 2 dimensional Ising magnets.

4. For long range forces, continuous symmetries can be broken in two dimensional non-zero temperature states (Pfister and Kunz).

The heuristic explanation of these 4 facts is that a 2 dimensional lattice system at non-zero finite temperature is a good approximation to an (ultraviolet cut-off) 2 dimensional Wightman theory in the Euclidean region. Thus, spontaneous breakdown of continuous symmetries is impossible, whereas there is no difficulty in 3 dimensions (explaining fact 2) or discrete symmetries (fact 3). Fact 4 is explained because the proof of Goldstone's theorem fails if there are long range forces: the presence of long range forces may give a mass to the otherwise massless Goldstone boson. How to make use of this to avoid massless particles in a relativistic theory is the subject of the next section.

4. THE HIGGS MECHANISM

The way to introduce long range forces in a relativistically covariant theory, and so give the Goldstone particle a mass, was solved by Higgs - we couple the system to a vector gauge field. Such theories are more useful since they are not plagued by massless particles not found in Nature.

A massless vector gauge field, A^μ, has free Lagrangian

$$\mathcal{L}_A = -\frac{1}{2} F^{\mu\nu}(\partial_\mu A_\nu - \partial_\nu A_\mu) + \frac{1}{4} F_{\mu\nu}F^{\mu\nu} \tag{4.1}$$

This is invariant under the gauge group

$$A^\mu \rightarrow A^\mu + \partial_\mu \Lambda \tag{4.2}$$

If there were a mass term $\frac{1}{2}m^2 A_\mu A^\mu$, in (4.1), it would not be invariant under this gauge group. Hence the proverb or saying: gauge fields have zero mass. This has been thought to be a difficulty of gauge theories of the Yang-Mill type; not enough zero-mass particles actually appear in Nature. Schwinger remarked that electrodynamics with zero mass fermions in two-dimensional space-time is exactly soluble; it is a gauge theory and the photon of the solution has a mass. This model is not very convincing, since in two-dimensions, A^μ has no dynamical degrees of freedom as it has no transverse components. Higgs [20],[22] pointed out that the mass of the gauge field need not be zero if the corresponding gauge symmetry (of the first type) is spontaneously broken, and that the presence of the (long range) gauge field avoids the Goldstone bosons of the spontaneously broken symmetry. Each concept helps the other avoid its zero mass difficulties. How this works is seen in the Brout-Englert model [16]. Consider the massless field given by (1.3)

$$\mathcal{L}_\phi(x) = \frac{1}{2} \chi^\mu(x)\chi_\mu(x) + \chi^\mu \partial_\mu \phi \tag{4.3}$$

We couple this "minimally" to a massless vector gauge field A^μ, of which the free Lagrangian is (4.1). "Minimal coupling" here means replace ∂_μ by $\partial_\mu + gA_\mu$ in $\mathcal{L}_\phi$. This rule leads to an interaction that depends on the form of $\mathcal{L}_\phi$: (4.3) leads to a different minimal coupling for ϕ from the usual free Lagrangian $\frac{1}{2}\partial_\mu\phi\partial^\mu\phi$. The advantage of (4.3) is that χ^μ does not contain ∂_μ explicitly, so that $\mathcal{L}_\phi$ is linear in ∂_μ. The interacting Lagrangian is thus

$$\mathcal{L}(g) = -\frac{1}{2} F^{\mu\nu}(\partial_\mu A_\nu - \partial_\nu A_\mu) + \frac{1}{4} F_{\mu\nu}F^{\mu\nu} + \chi^\mu \partial_\mu \phi$$
$$+ \frac{1}{2} \chi^\mu \chi_\mu + g\chi^\mu A_\mu \tag{4.3}$$

which is quadratic and therefore solvable. Our $\mathcal{L}(g)$ is gauge invariant under

$$\phi(x) \to \phi(x) + g\eta(x)$$
$$A_\mu(x) \to A_\mu(x) - \partial_\mu \eta \tag{4.4}$$

achieved by adding the coupling $g\chi^\mu A_\mu = j^\mu A_\mu$ to the free Lagrangians. This coupling appears to be local, but since A_μ is not a Wightman field (does not commute at space-like separation) it is a miracle that the theory is local.

We have to remove the arbitrariness of the solution due to the gauge freedom, (4.2). In the Coulomb gauge we do this by imposing the (non-covariant) transversality condition $\partial_k A^k = o$ $(k = 1,2,3)$, by adding $-C(x)\partial_k A^k(x)$ to $\mathcal{L}(x)$. $C(x)$ is a Lagrange multiplier (one for each x). The field equations are then

$$F^{\mu\nu} = \partial^\mu A^\nu - \partial^\nu A^\mu$$
$$\partial_\mu F^{\mu\nu} = - g\chi^\nu + \delta^\nu_k \partial^k C$$
$$\partial_k A^k = o \tag{4.5}$$
$$\chi_\mu = - \partial_\mu \phi - gA_\mu$$
$$\partial_\mu \chi^\mu = o$$

The dynamical variables are (ϕ, χ^o) and the two transverse components of A_k and their conjugates F^{ok}. The others are dependent variables

$$- \nabla^2 A^o = \partial_k F^{ok} = g\chi^o$$

Eliminating these, we get the dynamical equations

$$\partial_o A^T_k = F^T_{ok} \qquad \partial_o F^{okT} = - \nabla^2 A^{kT} + g^2 A^{kT}$$
$$\partial_o \phi = - \chi_o + g^2 (\nabla^2)^{-1} \chi_o \quad , \quad \partial_o \chi^o = \nabla^2 \phi$$

It follows that A_k^T, F_{ok}^T and χ^o satisfy the free wave-equation of mass g, and for the new variable

$$V_k = A_k + g^{-1}\partial_k\phi \tag{4.6}$$

we get the vector of mass g (three components). For ϕ, a non-local field, we get $(-\partial^2 + g^2)\partial_\mu\phi = o$, containing only $\partial_\mu\phi$. We may eliminate $\chi_\mu = -gV_\mu$ from the Lagrangian, which becomes

$$\mathcal{L}(g) = -\frac{1}{2}F^{\mu\nu}(\partial_\mu V_\nu - \partial_\nu V_\mu) + \frac{1}{4}F^{\mu\nu}F_{\mu\nu} - \frac{1}{2}g^2V^\mu V_\mu$$

$$- C\partial_k(V^k - g^{-1}\partial^k\phi) \tag{4.7}$$

It has three degrees of freedom, the V's and their conjugates F^{ok}. The theory has no massless bosons and the symmetry has disappeared from these gauge invariant fields. Let us examine the steps in the Goldstone theorem. The symmetry of the original Lagrangian $\phi \to \phi + \eta$ is generated by the conserved current $\chi^\mu = \frac{\delta\mathcal{L}}{\delta(\partial_\mu\phi)}$.
The proof starts with

$$f^\mu(x) = -ig\langle[\chi^\mu(x),\phi(o)]\rangle$$

and using the CCR, we get

$$\int d^3x f^o(\underline{x},o) = g \neq o$$

If this were true at all times, we would get the Goldstone theorem, and ϕ would have zero mass. However, as $\nabla^2\phi = g \operatorname{div}\vec{V}$,

$$f^\mu(x) = ig^3\langle[V^\mu(x),\nabla^{-2}\partial_k V^k]\rangle$$

and as

$$[V^\mu(x),V_\nu(y)] = (\delta^\mu_\nu - g^{-2}\partial^\mu\partial_\nu)\Delta(x-y;g^2)$$

we see that, while f^o has the causal structure

$$f_o(x) = g\partial_o\Delta\ (x,g^2)$$

the $\vec{f}$ has not:

$$f_k(x) = g(-g^2+\Delta^{-1})\partial_k\Delta(x,g^2)$$

Hence we can no longer throw away surface terms, and the space integral of f^o is time-dependent:

$$Q = \int d^3xf^o(\underline{x},t) = \cos gt$$

The field C is not a dynamical variable; it obeys $\nabla^2C=o$ and has no conjugate.

In the Lorentz gauge, best for the Euclidean version of this theory, we add the gauge fixing term $-G\partial_\mu A^\mu$, to get the equations of motion

$$\begin{aligned}
&\partial_\mu F^{\mu\nu} = -g\chi^\mu + \partial^\nu G && \text{(a)} \\
&\partial_\mu A^\mu = o && \text{(b)} \\
&F_{\mu\nu} = \partial_\nu A_\nu - \partial_\nu A_\mu && \text{(c)} \qquad (4.8)\\
&\chi_\mu = -\partial_\mu\phi - gA_\mu && \text{(d)} \\
&\partial_\mu\chi^\mu = o && \text{(e)}
\end{aligned}$$

It follows from (a) and (e) that $\Box\, G = o$: G is the Goldstone boson. The field ϕ, initially massless, has acquired a mass, as has A_k; combined, they form the three componenets of V_k.

This model is unrealistic, not least because the transverse components of A^μ acquire a mass. Actual electromagnetism contrives to maintain massless photons. The Higgs model is a non-linear model, in which, at least in perturbation theory, the gauge fields, and the Goldstone bosons, get a mass, by a mechanism similar to the one described here. There is a Lagrangian

$$\mathcal{L}_\phi = \frac{1}{2}\ (m_o^2\phi^*\ \phi + \nabla\phi^*\cdot\nabla\phi + \pi^*\pi)+\lambda(\phi^*\phi)^2 \qquad (4.9)$$

which we would expect to have solutions in which $\langle\underline{\Psi}_o,\phi_1(x)\underline{\Psi}_o\rangle = \eta \neq o$, where $\phi = \phi_1 + i\phi_2$. This provides the symmetry, SO(2), that is spontaneously broken, and gives us multiple vacua. In accordance with our philosophy, SO(2) must be a gauge group, and the observables are the SO(2)-invariant elements of the canonical algebra. The Goldstone bosons are still there and are, presumably, observable states.

We now couple $\mathcal{L}_\phi$ of (4.9) to a massless gauge field with free Lagrangian (4.1), using the minimal coupling notion. The resulting theory, checked by Higgs to lowest order, has no massless particles. As before, the symmetry group is trivial in that it is the identity on all observables; there is no trace of its breakdown or the massless Goldstone bosons. You might ask, what is the point of considering a spontaneously broken symmetry that goes away in its observable effects. These ideas are very important ("they have changed physics" - Zimmermann) because

1) Gauge theories are more renormalizable than expected by simple power counting in the Coulomb gauge.
2) The manifestly covariant form, especially the Euclidean theory, needs quantities like A_μ.
3) The several vacua enable us to construct soliton states, at least in two dimensions.

How 3) works was the subject of my 1973 Schladming lectures. Since then there have been important technical advances, mainly by Fröhlich [25].

5. QUANTUM SOLITONS AND COHOMOLOGY

Soliton has come mean various things in classical wave theory. For example given the Hamiltonian density, with V(O) = O, and

$$H(\underline{x}) = \frac{1}{2}(\dot{\phi}^2(\underline{x},o) + \nabla\phi\cdot\nabla\phi) + V(\phi) \qquad (5.1)$$

we may seek <u>non-dissipative</u> solutions i.e. solutions such that

$$\sup_{\underline{x}} |\varphi(\underline{x},t)| \geq c > o \quad \text{for all time } t.$$

Alternatively, solitons may be thought of as solutions for which $\varphi(\underline{x},o)$ is not zero at spatial ∞, but which have finite energy. As we see from (5.1), to have finite energy, a solution "must" approach, at spatial ∞, a zero of $V(\phi)$, and must be slowly varying there, in order for $|\nabla\phi|^2$ to be square integrable. According to the "Goldstone picture", we may expect the corresponding quantum field theory to exhibit several vacuum states if $V(\phi)$ has several minima as a function of ϕ. The states invariant under time evolution are then, approximately, states in which $\phi(x) = \xi_i$, where $\xi_i, i = 1,2,\ldots$ are zeroes of V. Some people take solitons to be stationary solutions that are not space-translatio invariant; others do not require stationarity, but require rather that the kinetic energy density is finite. All these ideas should be borne in mind when attempting to quantize the theory described by (5.1).

In order to avoid dissipative solutions, it would seem that, as $|\underline{x}| \to \infty$ in different directions, $(\underline{x})$ should converge to different zeroes of V. Otherwise our solutions will not differ much from one of the vacuum solutions $\phi(x) = \xi_i$, and so could be thought of as being a "normal" or non-soliton solution, converging to ξ_i as $t \to \infty$.

In the quantum version, the ξ_i should be linked by a symmetry group, so that this group can be taken as the gauge group, and multiple vacua avoided, in the sense that the vacua $\phi = \xi_i$, $i = 1,2,\ldots$ become physically indistinguishable.

Different solitons have been distinguished by "topological" quantum numbers. In the quantum version, naturally, the different vacua are to define inequivalent representations of the algebra $\mathfrak{A}$. Thus, this quantum number becomes a label for inequivalent representations, i.e. it is like a Casimir operator for Lie algebras. It is this property, rather than anything topological, that is the more fundamental.

Let us study in detail a free field, given by a wave-equation

$$(\Box + m^2)\phi_\alpha(\underline{x},t) = 0 \tag{5.2}$$

$$D_\alpha\phi_\alpha(\underline{x},t) = 0 \tag{5.3}$$

where (5.3) is a supplementary condition eliminating the unwanted spin components. Suppose that the solutions realize a representation $U_1 = \bigoplus_s [m,s]$ of $\mathcal{P}$, and that second quantization is carried out to give a local c^*-algebra theory, as in §2. We recall that the Poincaré group $\mathcal{P}$ acts on the algebra of Weyl operators $W(\phi)$ by automorphisms

$$\tau_L W(\phi) = W(\phi_L), \quad L \in \mathcal{P} \tag{5.4}$$

where $\phi \in \mathcal{H}_1$, the one-particle space on which U_1 acts. The canonical structure in $\mathcal{H}_1$ is determined by the real subspace $\mathcal{H}$ of time-reversal invariant vectors. These correspond to the 'position variables', the q's of a canonical theory. We shall denote by $W(\phi)$ the usual relativistic Fock representation of the Weyl operators. They obey the Segal-Weyl relations (2.7). We recall the definition of a <u>covariant representation</u> π of $\mathcal{O}\!\mathcal{L}$, namely, one in which the automorphism group τ_L is continuously represented by $U_\pi(L)$, giving positive energy.

Fröhlich has proved that if $m > o$, then the Fock representation is the only covariant representation. However, interesting things (like spontaneously broken symmetries) happen if $m = o$. We might ask, what are <u>all</u> covariant representations of this system? This is quite a hard question, but we are able to classify all covariant representations of the special class known as <u>displaced Fock representations</u>.

<u>Definition</u>. Let $\mathcal{H}_1 = \mathcal{H}_{1r} + i\,\mathcal{H}_{1r}$ be a one-particle space, and let π_F be the Fock representation on $\mathcal{H}_1$. Let $\phi^\times \in \mathcal{H}_1^\times$ = algebraic dual of $\mathcal{H}_1$. Then the representation defined by

$$\pi_{\phi^\times}\,(W(\phi)) = e^{i\mathrm{Im}\langle\phi^\times,\phi\rangle} W_F(\phi)$$

is called the displaced Fock representation determined by $\phi^\times$. Using $W(\phi) = e^{i\{\hat{\phi},\phi\}}$, we see that, formally

$$\pi_{\phi^\times} W(\phi) = W(\phi + \phi^\times)$$

where $\phi^\times$ is the solution of (5.2), (5.3) determined by $\phi^\times$ i.e. the generalized functions, $\phi^\times(\underline{x},0)$, $\phi^\times(x,o)$ that, smeared with $\phi(\underline{x},o), -\phi(\underline{x},o)$ gives the number $\mathrm{Im}\,\langle\phi^\times,\phi\rangle$. It is known that $\pi_{\phi^\times}$ is equivalent to π_F if and only if $\phi \to \text{Im} <\phi^\times, \phi>$ is a continuous linear functional on $\mathfrak{M}_o$, the chosen test-function space for the canonical system (Shale's theorem, in the form given by Manuceau). To get a new representation, we must therefore choose $\phi^\times \notin \mathfrak{M}^* = \mathcal{H}$, where $\mathcal{H}^*$ is the topological dual (= set of continuous linear functionals). Let us do this.

To ensure that the representation $\pi_{\phi^\times}$ is covariant, we must have that the representation $W(\phi) \to \pi_{\phi^\times}(W(\phi))$ and the representation $W(\phi) \to \pi_{\phi^\times} W(U_1(L)\phi)$ are equivalent. This is the same as asking that the map

$$\pi_F(W(\phi)) \to e^{i\, \text{Im}<\phi^\times, U_1(L)\phi - \phi>} \pi_F\, W(\phi_L)$$

should be implemented by a unitary. By the Shale criterion, this requires that $\phi \to \text{Im} <\phi^\times, \phi_L - \phi>$ should be continuous in ϕ, i.e. that

$$U_1^\times(L)\phi^\times - \phi^\times \in \mathcal{H}_1$$

where $U_1^\times$ is the dual action of $\mathcal{P}$ on $\mathcal{H}_1^\times$. We remark that the map $\psi : L \to U_1^\times(L)\phi^\times - \phi^\times$ obeys the <u>cocycle condition</u> relative to the action of U an $\mathcal{H}$, namely

$$U_1^*(M)\psi(L) = \psi(LM) - \psi(M) \tag{5.5}$$

It can be shown that every continuous map $\psi, \mathcal{P} \to \mathcal{H}_1$, obeying (5.5) can be written

$$\psi(L) = \phi^\times - U_1^\times(L)\phi^\times$$

for some $\phi^\times \in \mathcal{H}^\times$. If $\phi^\times \in \mathcal{H}, \psi(L)$ is called a coboundary. Then we have

<u>Theorem</u> There is a 1:1 correspondence between continuous finite energy cocycles ψ_L of the representation U_1 of $\mathcal{P}$ on $\mathcal{H}_1$, and the covariant representations $\pi_{\phi^\times}$ of $\mathfrak{A}$. In this correspondence, the operator $V_{\phi^\times}(L)$ defined on states $W(\phi)\underline{\Psi}_o$ by

$$V_{\phi^\times}(L) W_F(\phi)\underline{\Psi}_o = e^{i/2\, \text{Im}<\phi, \psi_L>} W(\phi_L - \psi_{L^{-1}})\underline{\Psi}_o$$

implement the Poincaré automorphism τ_L in the representation $\pi_{\phi^\times}$, and the $V_{\phi^\times}(L)$ is a continuous multiplier representation of $\mathcal{P}$ with multiplier $\omega(L,M)=e^{i\mathrm{Im}\langle\psi_L,\psi_{M^{-1}}\rangle}$. Also, $V_{\phi^\times}$ is ∞-divisible.

This theorem is not hard. The details will appear in a joint paper with J. Wright. See also Roepstorff [26].

The state defined by

$$W(\phi) \to \langle\underline{\Psi}_o, \pi_{\phi^\times}(\phi)\underline{\Psi}_o\rangle$$

has the property that it gives, for the expectation of the quantized field $\hat{\phi}$, the classical wave $\phi^\times(\underline{x},t)$. Thus, this state is the coherent state corresponding to $\phi^\times$. The solution $\phi^\times$ to the wave-equation does not lie in $\mathcal{H}_1$ if $\pi_{\phi^\times}$ is not equivalent to π_F. This turns out to be caused by the fact that $\phi^\times$ is not small at ∞. However, $\phi^\times$ is a wave of finite energy. Thus, $\phi^\times$ looks a bit like a soliton. Its shape is not determined locally, but its asymptotic behaviour is determined by its cohomology class, labelled by conserved quantum numbers.

It is known that if $m > o$, there are no non-trivial cocycles of any representation of $\mathcal{P}$. For $m = o$ in two space-time dimensions, the inreducible representation has exactly $\mathbb{R}^2$ cocycles modulo coboundaries. These are generated by real multiples of the two (soliton) solutions with Cauchy data

$$\dot{\phi}_1^\times(\underline{x},o) = o \quad , \quad \phi_1^\times(\underline{x},o) = \theta(\underline{x})$$

$$\dot{\phi}_2^\times(\underline{x},o) = \Delta(\underline{x}), \quad \phi_2^\times(\underline{x},o) = o$$

where $\theta(\underline{x})$, $\Delta(\underline{x})$ are smooth approximations to the step function and δ-function respectively. It can be shown that the charged sectors described by $\phi^\times = q_5\phi_1^\times + q\phi_2^\times$ are the states of the massless Thirring model of change q_5 and q. Clearly, $q_5 = \int_{-\infty}^{\infty} \nabla\phi^\times dx$ and $q = \int \dot{\phi}^\times dx$ are the axial charge and charge, respectively. The charge q_5 coincides with the so-called topological charge; but its origin is seen here to be cohomoligical.

REFERENCES

For §1

[1] R.F. Streater, Spontaneous Breakdown of Symmetry in Axiomatic Theory, Proc. Roy. Soc. A287, 51o - 8 (1965).

For §2

[2] R.F. Streater and I.F. Wilde, Fermion States of a Boson Field, Nucl. Phys. B24, 561 - 75 (197o).

For §3

[3] J. Goldstone, A. Salam and S. Weinberg, Broken Symmetries, Phys. Rev. 127, 965 (1962).

[4] H. Ezawa and J.A. Swieca, Spontaneous Breakdown of Symmetries and Zero-mass States, Commun. Math. Phys. 5, 33o - 336 (1967).

[5] R. Jost and H. Lehmann, Integraldarstellung einer causalen Kommutatoren', Nuovo Cimento 5, 1598 - 61o (1957).

[6] S. Coleman, There are no Goldstone Bosons in Two Dimensions, Commun. Math. Phys. 31, 259 - 264 (1973).

[7] A.S. Wightman, Cargèse Lectures, 1964, Gordon and Breach, N.Y. 1966, Introduction to some Aspects of the Dynamics of Quantized Fields.

[8] E. Gal-Ezer, Spontaneous Breakdown in Two-dimensional Space-time. Commun. Math. Phys. 44, 191 - 196 (1975).

[9] R.F. Streater, The Heisenberg Ferromagnet as a Quantum Field Theory, Commun. Math. Phys. 6, 233 - 247 (1967).

[1o] N.D. Mermin, Absence of Ordering in Certain Classical Systems, J. Math. Phys. 8, 1o61 - 1o64 (1967).
N.D. Mermin and H. Wagner, Absence of Ferromagnetism or Antiferromagnetism in One-or Two-Dimensional Isotropic Heisenberg Models, Phys. Rec. Lett. 17, 1133 (1966).

[11] F.J. Dyson, E. Lieb and B. Simon, to be published.

[12] Ch.-E. Pfister and H. Kunz, First Order Phase Transition in the Plane Rotator Ferromagnetic Model in Two Dimensions, Commun. Math. Phys. 46, 245 - 251 (1976).

[13] R. Lange, Goldstone Theorem in Non-Relativistic Theories, Phys. Rev. Lett. 14, 3 - 6 (1965).

[14] R.F. Streater, Spontaneous Breakdown of Symmetry, in Mathematical Theory of Elementary Particles, MIT Press, 1966. Ed. R. Goodman and I.E. Segal.

[15] D. Kastler, D.W. Robinson and J.A. Swieca, Conserved Currents and Associated Symmetries: Goldstone's Theorem, Commun. Math. Phys. 2, 1o8 - 2o (1966).

For §4

[16] R. Brout and F. Englert, Broken Symmetry and the Mass of Gauge Vector Mesons, Phys. Rev. Lett. 13, 321 - 323 (1964).

[17] D.G. Boulware and W. Gilbert, Connection between Gauge Invariance and Mass, Phys. Rev. 126, 1563 - 1567 (1962).

[18] G. Guralnik, C.R. Hagen and T.W.B. Kibble, Global Conservation Laws and Massless Particles, Phys. Rev. Lett. 13, 585 (1964).

[19] G. Guralnik, C.R. Hagen and T.W.B. Kibble, Advances in Elementary Particle Physics, Vol. II, Interscience 1968. Ed. R.L. Cool and R.E. Marshak.

[20] P.W. Higgs, Broken Symmetries and the Masses of Gauge Bosons, Phys. Rev. Lett. 13, 508 - 509 (1964). See also Phys. Lett. 12, 132 (1964).

[21] R.F. Streater, Broken Symmetries and the Goldstone Theorem in: High Energy Physics and Elementary Particles, Ed. A. Salam, International Atomic Energy Agency, Vienna, 1965.

[22] P.W. Higgs, Spontaneous Symmetry Breakdown without Massless Bosons, Phys. Rev. (2), 145, 1156 - 1163 (1966).

For §5

[23] R.F. Streater and J. Wright, Solitons are Cocycles, to be published.

[24] J.-L. Bonnard and R.F. Streater, Local Gauge Models Predicting their own Superselection Rules, Helv. Phys. Acta, 49, 259 - 267 (1976).

[25] J. Fröhlich, New Superselection Sectors ("Soliton States") in Two-Dimensional Bose Quantum Field Models, Commun. Math. Phys. 47, 269 - 310 (1976).

[26] G. Roepstorff, Coherent Photon States and Spectral Condition, Commun. Math. Phys. 19, 301 - 314, (1970).

SHORT INTRODUCTION TO NONSTANDARD ANALYSIS AND ITS PHYSICAL APPLICATIONS

Jan Tarski

University of Bielefeld

Department of Theoretical Physics

ABSTRACT

Rudiments of nonstandard analysis are presented. Some previous applications to quantum field theory models and to the study of the thermodynamic limit are discussed.

1. PRELIMINARY REMARKS

The subject matter of nonstandard (n.s.) analysis can be described as the analysis of an extended number system, which includes infinite and infinitesimal quantities. This subject, like other developments in analysis, can sometimes streamline the handling of limiting processes. Physicists might therefore attempt applications to quantum field theory and to many body theory. Quantum field theory in particular, where infinities arise in natural ways, might profit from such applications (cf.[1,2]).

This article is organized as follows. In §§2-5 we present the barest essentials concerning nonstandard extensions, and in §6 we comment on nonstandard existence proofs for thermodynamic limits. Next, §§7-8 contain the rudiments of n.s. Hilbert spaces and associated linear operators, while in §§9-1o we apply these constructions to quantum field theory models.

The applications considered in §§6 and 9-1o have been the subject of previous articles. However, our discussion supplements (rather than summarizes) the published material. E.g., in §6 we prove the equivalence of two ways of determining the thermodynamic limit.

The last sections show some poins of contact between n.s. Hilbert spaces on one hand, and rigged Hilbert spaces, the HS construction, and infinite tensor products on the other. This circumstance may suggest a remarkable versatility of n.s. techniques. However, for the present we cannot say that the n.s. techniques offer an advantage over alternative constructions.

These notes constitute an expanded version of a seminar given at the Bielefeld Summer Institute. This seminar was intended as an introduction to a seminar of A. Ostebee, who was not able to prepare a text of his talk. However, most of the material that he presented is readily available [3], and our §6 is devoted to the same subject.

We make no attempt to give here an extensive bibliography on n.s. analysis, but we note the following: the summaries intended for physicists in [2] and [4], also a systematic text [5], and a thorough introduction [6].

The author thanks Dr.A.Ostebee for a useful discussion. He expresses his appreciation to Profs.H.Satz and L.Streit, who organized the Summer Institute. He gratefully acknowledges the support of Deutsche Forschungsgemeinschaft.

2. ULTRAFILTERS AND THE HYPERREAL LINE

We give now a construction of nonstandard numbers. Such a number is an equivalence class of sequences: $(a_1, a_2, \ldots)$, where each $a_j \in R^1$. Two sequences $(a_1, \ldots)$ and $(b_1, \ldots)$ are equivalent if the indices for which the entries agree form a subset of the positive integers N belonging to a selected family $\mathcal{U}$:

$$(a_1, \ldots) \equiv (b_1, \ldots) \Longleftrightarrow \{j : a_j = b_j\} \in \mathcal{U}. \qquad (1)$$

(Note that nothing has been said about convergence.) The family $\mathcal{U}$ must be a free ultrafilter over N, i.e. must satisfy, for all $\underline{X}$, $Y \subseteq N$:

(i) if $\underline{X} \in \mathcal{U}$ and $Y \supseteq \underline{X}$ then $Y \in \mathcal{U}$;
(ii) $N \in \mathcal{U}$, $\emptyset \notin \mathcal{U}$;
(iii) if $\underline{X} \in \mathcal{U}$ and $Y \in \mathcal{U}$ then $\underline{X} \cap Y \in \mathcal{U}$;
(iv) $X \in \mathcal{U}$ or $\underline{X}^c = N \setminus \underline{X} \in \mathcal{U}$;
(v) if $\underline{X}$ is a finite set then $\underline{X}^c \in \mathcal{U}$.

A family satisfying (i)-(iv) is, more generally, an ultrafilter. For this reason we listed (ii) separately, even though it is a consequence of (v). We now comment on these properties in turn.

The property (i) can be described as a filtering property, and then (ii) eliminates the two extreme cases: $\underline{X} \in \mathcal{U}$ for $\forall \underline{X}$, or $\underline{X} \notin \mathcal{U}$ for $\forall \underline{X}$. Next, (iii) ensures that $\equiv$ is transitive, i.e.

$$(a_1,\ldots) \equiv (b_1,\ldots) \text{ and } (b_1,\ldots) \equiv (c_1,\ldots) \Rightarrow (a_1,\ldots) \equiv (c_1,\ldots) , \tag{2}$$

and (iv) is a maximality property.

A trivial example of an ultrafilter [satisfying (i)-(iv)] is the following: Select any $m \in N$, and let

$$\underline{X} \in \mathcal{U} \iff m \in \underline{X}. \tag{3}$$

In this case the sequence $(a_1,\ldots)$ would be fully characterized up to equivalence by the single entry a_m, and the set of equivalence classes of sequences could be identified with the real line. The property (v) excludes this trivial possibility.

There is no simple way to exhibit a free ultrafilter $\mathcal{U}$ over N. The existence of such a $\mathcal{U}$, however, is a consequence of the axiom of choice. In fact, there exists an infinity of them. We suppose that a particular $\mathcal{U}$ has been chosen, and it will be held fixed for the remainder of the article.

We will denote the equivalence class of a sequence by square brackets. In case of real numbers, the set of such equivalence classes is the <u>hyperreal line</u>, denoted by $^*R^1$. Thus

$$^*R^1 = \{[(a_1,a_2,\ldots)] : a_j \in R^1 \text{ for } \forall j \in N\}. \tag{4}$$

We will identify $a \in R^1$ with $[(a,a,..)] \in {}^*R^1$, and then

$$R^1 \subset {}^*R^1. \tag{5}$$

Elements of $^*R^1$ of the form $[(a,a,\ldots)]$ will be called <u>standard</u> elements.

More generally, if $\mathcal{S}$ is any set, we define

$$^*\mathcal{S} = \{[(s_1,s_2,\ldots)] : s_j \in \mathcal{S} \text{ for } \forall j\}. \tag{6}$$

Nonstandard extensions like this one will sometimes be denoted by prefixing a star, as here. However, the star will often be omitted if the meaning is clear.

Finally, we remark that other nonstandard extensions of R^1, etc., are possible. E.g. one can suppose that the index j runs over a continuum rather than over a discrete set. The extension described above is generally sufficient.

3. A FEW PROPERTIES OF HYPERREAL NUMBERS

While the basic arithmetical operations on real numbers can be readily adapted to other systems, the ordering properties are more difficult to adapt. For the hyperreal line this is possible, and the property (iv) of ultrafilters is crucial.

Let $a=[(a_1,a_2,\ldots)]$ and $b=[(b_1,b_2,\ldots)]$. We consider

$$\underline{X}_{\leq} = \{j : a_j \leq b_j\} , \quad \underline{X}_{>} = \{j : a_j > b_j\} ; \tag{7}$$

and the analogous sets $\underline{X}_{<}$, $\underline{X}_{\geq}$, and $\underline{X}_{=}$. But

$$\underline{X}_{\leq} = \underline{X}_{>}{}^{c} = N \setminus \underline{X}_{>} , \tag{8}$$

and so

$$\underline{X}_{\leq} \in \mathcal{U} \quad \text{or} \quad \underline{X}_{>} \in \mathcal{U} \quad \text{(not both)}. \tag{9}$$

In the first case we say $a \leq b$, and in the second $a > b$. It is straightforward to verify that this definition is independent of the representatives $(a_1,\ldots)$ and $(b_1,\ldots)$. Similarly we define $a \geq b$ and $a < b$, and readily conclude:

Proposition 1. Exactly one of the following holds:

$$a < b \quad \text{or} \quad a = b \quad \text{or} \quad a > b. \tag{10}$$

We define next $|a|$ by

$$|a| = a \quad \text{if} \quad a \geq 0, \quad |a| = -a \quad \text{if} \quad a \leq 0 , \tag{11}$$

where $-a = [(-a_1, -a_2,\ldots)]$, and we say that
a is finite if $|a| < u$ for some $u \in R^1$,
a is infinitesimal if $|a| < \varepsilon$ for $\forall \varepsilon \in R^1, \varepsilon > 0$
a is infinite if $|a| > v$ for $\forall v \in R^1$.
Thus, the property (v) of free ultrafilters implies that

$$[(1,\tfrac{1}{2},\tfrac{1}{3},\ldots)] \text{ is infinitesimal,} \tag{12a}$$

$$[(1,2,3,\ldots)] \text{ is infinite.} \tag{12b}$$

Proposition 2. If a is finite, then $\exists$ unique ${}^{\circ}a \in R^1$ such that $a - {}^{\circ}a$ is infinitesimal.

The number ${}^{\circ}a$ is given by

$${}^{\circ}a = \mathrm{glb}_{u \in R^1}(u > a) = \mathrm{lub}_{v \in R^1}(v < a), \tag{13}$$

and it is trivial to verify that $a - {}^{o}a$ is infinitesimal.
We will write if a is finite,

$$^{o}a = \text{std}\ (a), \tag{14}$$

and we define the monad $\mu(b)$ (for any $b \in {}^{*}R^{1}$) and the relation $\simeq$ by

$$\mu(b) = \{b \in {}^{*}R^{1} : b - c \ \text{ is infinitesimal}\}, \tag{15}$$

$$b \simeq c \Longleftrightarrow b - c \ \text{ is infinitesimal.} \tag{16}$$

Now (12a) generalizes to: If $a_k \to a$ (in R^1) then

$$[(a_1, a_2, \ldots)] \in \mu(a) \ . \tag{17}$$

We next consider arithmetical operations. With a and b as before, we set

$$a + b = [(a_1 + b_1, \ a_2 + b_2, \ldots)] \ , \tag{18}$$

etc. The addition as here extended to $^{*}R^{1}$ remains commutative and associative. The other basic laws of arithmetic similarly remain valid (see also §4). In particular, if $a \neq o$ then $\exists$ unique a^{-1} such that $aa^{-1} = a^{-1}a = 1$.
One can similarly extend the definition of an arbitrary function $f: R^1 \to R^1$, namely

$$f(a) = [(f(a_1), \ f(a_2), \ldots)] \ . \tag{19}$$

Suppose now that $b \in R^1$, that $b_k \to b$, and that f is continuous at b. Then (17) implies that $f(\mu(b)) \subseteq \mu(f(b))$. The converse is also valid ([5], p.66), and so:

<u>Proposition 3.</u> f is continuous at $a \in R^1$ if and only if $f(\mu(a)) \subseteq \mu(f(a))$.
If the function f is differentiable at $a \in R^1$, then its derivative can be expressed in a form very similar to $df(b)/db\,\big|_{b=a}$. Proposition 3 leads directly to:

<u>Proposition 4.</u> f has the derivative $f'(a)$ at $a \in R^1$ if and only if for $\forall\, b \in \mu(a)$,

$$\text{std}\ \{[f(b) - f(a)] \ / \ \ (b-a)\} = f'(a) \ . \tag{2o}$$

4. REMARKS ON THE TRANSFER THEOREM

One naturally expects that various properties of real numbers extend to the hyperreals. However, in some respects the two systems must necessarily differ. It is important to characterize the two kinds of properties.

We saw in §3 how the basic arithmetical laws can be extended to the hyperreals by arguing "by components". The same holds for functional relations, inequalities, etc., and the <u>transfer theorem</u> asserts that properties which are called internal carry over to a nonstandard extension. To formulate this theorem in full generality requires some technical terminology from logic. We do not attempt to give such a formulation here, but confine ourselves to stating the following assertion, which is adapted from [6] :

> Let f be a function of k variables ($k<\infty$), $f:A \to R^1$, where $A \subseteq R^k$. Then f has a natural extension (which will usually be denoted by the same symbol), $f:{}^*A \to {}^*R^1$, and this extension satisfies:
> (i) If a system of real formulas is true of all reals then the natural extension of this system is true of all hyperreals.
> (ii) Let S_1 and S_2 be two systems of real formulas. If the implication $S_1 \Rightarrow S_2$ is valid, then the implication remains valid for the corresponding natural extensions of S_1 and S_2.

We give simple examples: for (i), $e^{x+y}=e^x e^y$, and for (ii), $x > 0 \Rightarrow \sqrt{x^2} = x$. These are valid for all reals, and hence for all hyperreals. We note that a formula may include assertions of existence, e.g. $\exists x \in A$, but then the extended formula must have $\exists x \in {}^*A$.

We also give two examples of properties that exhibit a distinction between R^1 and ${}^*R^1$.

(a) The Archimidean property is the following: If $p,q \in R^1$, $p,q > 0$, then $\exists k \in N$ such that $kp > q$. This does not extend to the case $p,q \in {}^*R^1$ if k remains restricted to N.

(b) The function std of §3 agrees with the identity on R^1, but is not the natural extension of the identity function on R^1. Hence (i) and (ii) do not apply to the function std.

We may use here the terminology introduced previously, and say that std is not an internal function, or, that it is an external function. Likewise the set N is an external subset of ${}^*N$ or of ${}^*R^1$.

5. COMPLEX NUMBERS, NONSTANDARD INTEGERS, AND SEQUENCES

The present section is included as background material for the remainder of the article. However, we note that n.s. analytic functions, defined on ${}^*C^1$, and n.s. arithmetic, a study of ${}^*N$, are subjects of independent interest. See [5] .

Consider the set $\{1,i\}$. We write

$$[(1,1,\ldots)] = {}^*1, \quad [(i,i,\ldots)] = {}^*i, \tag{21}$$

and property (iv) of ultrafilters tells us that

$$^*\{1,i\} = \{^*1, {}^*i\}. \tag{22}$$

Moreover, if $c_j = a_j+ib_j$ then

$$[(c_1,c_2,\ldots)]=[(a_1,a_2,\ldots)]+(^*i)[(b_1,b_2,\ldots)]. \tag{23}$$

These considerations can be adapted easily to finite sets and to finite-dimensional vector spaces, as follows. (Finiteness here is crucial.)

Proposition 5. (a) Let S be a finite set. Then $^*S$ is a set with the same number of elements, and can be identified with S .

(b) If $k \in N$ then $^*(R^k)$ can be identified naturally with $(^*R^1)^k$. Moreover, the space $^*C^1$ has the structure $^*R^1 + (^*i)^*R^1$.

In view of this proposition, we can readily extend various considerations from real n.s. to complex n.s. quantities.

We now recall that N is the set of positive integers. Thus $N \subset R$, and $^*N \subset {}^*R^1$. The following property of $^*N$ is basic, and depends on property (iv) of ultrafilters (like Propositions 1 and 5):

Proposition 6. Every finite element of $^*N$ is also an element of N.

The set of $^*N$ is of course larger than N (in fact, is non-denumerable), and includes infinite elements, called infinite integers.

Let $\{p_k\}$, $k \in N$, be a standard sequence. We may think of $\{p_k\}$ as a function $p:N\to C^1$, and this function then has a natural extension to $p:{}^*N\to{}^*C^1$. We obtain in this way a sequence $\{p_n\}$, $n \in {}^*N$, which constitutes an extension of the original one.

The following can be proved ([5], pp. 60-63):

Proposition 7. The standard sequence $\{p_k\}$, $k \in N$, has a limit (in C^1) if and only if $p_n \simeq p_m$ for all infinite n,m. Moreover, the following are equivalent: (i) this limit is $\overline{p}$,(ii) $p_n \in \mu(\overline{p})$ for all infinite n, (iii) $p_n \in \mu(\overline{p})$ for all infinite n of the form $[(k_1, k_2,\ldots)]$ where $k_1<k_2<\ldots$.

The condition (iii) was not included in loc.cit., but it will be useful in the sequel. The implication (iii) $\Rightarrow$ (i) can be proved by assuming that $\{p_k\}$ does not approach $\overline{p}$ and then constructing n which violates (iii).

6. REMARKS ON EXISTENCE PROOFS FOR THERMODYNAMIC LIMITS

Let us consider a system of $\bar{N}$ particles, which interact through the potential $U_{\bar{N}}(\underset{\sim}{r}_1,\ldots,\underset{\sim}{r}_{\bar{N}})$ and which are confined to a region $\Lambda \subset R^3$ with finite volume $v(\Lambda)$. Let T be the temperature of the system, and we introduce as usual

$$p=\bar{N}/v(\Lambda), \qquad \beta=1/kT, \qquad \lambda=\hbar\,(2\pi/m\beta)^{\frac{1}{2}}. \tag{24}$$

Then the free energy g per unit volume is given by

$$g(\beta,p,\Lambda) = v(\Lambda)^{-1} \log \mathbf{Z}(\beta,\bar{N},\Lambda), \tag{25a}$$

$$\mathbf{Z}(\beta,N,\Lambda)=(\lambda^{3\bar{N}}\bar{N}!)^{-1} \int_\Lambda d^3\underset{\sim}{r}_1 \ldots \int_\Lambda d^3\underset{\sim}{r}_{\bar{N}} \exp[-\beta U_{\bar{N}}(\underset{\sim}{r}_1,\ldots,\underset{\sim}{r}_{\bar{N}})]. \tag{25b}$$

We are interested in investigating the limit of g as $v(\Lambda)\to\infty$ with β and p held constant. The latter two variables will be suppressed from now on.

Such an investigation often involves an intricate interplay of several sequences. Typically, we first take an increasing sequence of cubes Γ_ν and determine $\lim_{\nu\to\infty} g(\Gamma_\nu)$. Then we take a more general sequence of domains Λ_ν, satisfying suitable restrictions, we enclose each Λ_ν in a cube Ω_ν, and pack the regions Λ_ν and $\Omega_\nu\setminus\Lambda_\nu$ with small cubes in order to obtain further estimates.

Refs.[3] describe an attempt to simplify the estimating by using n.s. notions. The first of these articles deals with the case of strongly tempered potentials, i.e. those which satisfy

$$U_{N_1+N_2}(\underset{\sim}{x}_1,\ldots,\underset{\sim}{x}_{N_1};\ \underset{\sim}{y}_1,\ldots,\underset{\sim}{y}_{N_2})-U_{N_1}(\underset{\sim}{x}_1,\ldots)-U_{N_2}(\underset{\sim}{y}_1,\ldots) \le 0 \tag{26}$$

whenever the coordinates satisfy $|x_k-y_\ell| \ge R$, for a fixed R. For this case both the standard proof that $\lim g(\Lambda_\nu)$ exists [7] and the corresponding nonstandard proof are quite simple. We refer to the cited works for details, and also for additional conditions on the potential and on the domains. However, we should like to comment here on the passage from the nonstandard arguments in [3] to the desired conclusion, that $\lim g(\Lambda_\nu)$ exists. In our opinion, this point was not made clear in [3].

As in loc.cit., we take a suitable sequence of domains Λ_ν, and we let

$$^*\Lambda = [(\Lambda_1\Lambda_2,\ldots)] \quad \text{and} \quad \bar{g} = \text{std } g(^*\Lambda). \tag{27}$$

It is not knwon at this stage whether lim $g(\Lambda_\nu)$ exists. However, let us suppose that for every subsequence of these domains one also has

$$\bar{g} = \text{std } g([(\Lambda_{k_1},\Lambda_{k_2},\ldots)]) \text{ (where } k_1 < k_2<\ldots). \tag{28}$$

Then by Proposition 7, this condition is equivalent to

$$\bar{g} = \lim g(\Lambda_\nu).$$

Now, the conditions on sequences $\{\Lambda_\nu\}$ translate immediately to conditions on nonstandard domains like $^*\Lambda$. But if these conditions are fulfilled for $^*\Lambda$, then they are also fulfilled for every n.s. domain such as in (28). Consequently, the following are equivalent:

$$\lim_\nu g(\Lambda_\nu) = \bar{g} \quad \text{for every suitable } \{\Lambda_\nu\}, \tag{30a}$$

$$\text{std } g(^*\Lambda) = \bar{g} \quad \text{for every suitable } ^*\Lambda . \tag{30b}$$

In [3] the condition (30b) is taken as the goal of each investigation. It is claimed there that by adopting this condition and by using n.s. notions, one can clarify the structure of proofs in some cases, e.g. for Coulomb systems.

In §8 we compare a standard proof with its nonstandard counterpart in a different context.

7. NONSTANDARD HILBERT SPACES

Let $\mathcal{H}$ be a (standard) infinite-dimensional, separable Hilbert space over the complex numbers, with basis vectors $e_1,e_2,\ldots$ satisfying $\langle e_k, e_\ell\rangle = \delta_{k\ell}$. We construct a n.s. extension of this structure, with $^*\mathcal{H}$ the new space. The sequence $\{e_k\}$, $k \in N$, then extends to $\{e_n\}$, $n \in {}^*N$, with each $e_n \in {}^*\mathcal{H}$. It is readily seen that

$$\langle e_n, e_m\rangle = \delta_{nm} \quad \text{for} \quad n,m \in {}^*N . \tag{31}$$

Thus the e_n form an orthonormal set of $^*\mathcal{H}$. They form indeed the whole basis, in view of the following:

Proposition 8. Let $v \in {}^*\mathcal{H}$. Then

$$v = \sum_{n=1,n\in {}^*N}^{\infty} \langle e_n,v\rangle\, e_n , \tag{32}$$

in the sense that given $\varepsilon \in {}^*R^1$, $\varepsilon > o$, $\exists\, m^o \in {}^*N$ such that

$$||v- \sum_{n=1}^{m} \langle e_n,v\rangle \; e_n||< \varepsilon \quad \text{for all } m \geq m^o \; . \qquad (33)$$

The last relation follows by transfer from $\mathcal{H}$. However, the sums here are non-denumerable but discrete, of quantities which might be infinitesimal. The ideas are becoming a bit intricate, so we give a few details. Let

$$\varepsilon = [(\varepsilon_1,\varepsilon_2,\ldots)] \quad , \qquad v = [(v_1,v_1,\ldots)] \; . \qquad (34)$$

Choose an index ℓ, and let $m_\ell \in N$ be such that for all $k > m_\ell$,

$$||v_\ell - \sum_{j=1,j\in N}^{k} \langle e_j,v_\ell\rangle \; e_j|| < \varepsilon_\ell \; . \qquad (35)$$

Then we may take $m^o = [(m_1,m_2,\ldots)]$. The summation $\sum_n$ has of course the meaning

$$[(\textstyle\sum_{n_1}, \sum_{n_2},\ldots)] \; . \qquad (36)$$

The proposition now implies that

$$||v||^2 = \langle v,v\rangle \;=\; \sum_{n\in {}^*N} |\langle e_n,v\rangle|^2 \quad . \qquad (37)$$

We will say that v is infinitesimal, finite, or infinite, depending on the value of $||v||$. We set, as in (15),

$$\mu(v) = \{u \in {}^*\mathcal{H} : u - v \;\text{ is infinitesimal}\} \; . \qquad (38)$$

For infinite-dimensional spaces a new concept is worth introducing. A vector v is called <u>near-standard</u> if $\exists\, v^o \in \mathcal{H}$ such that $v \in \mu(v^o)$. E.g., the vectors e_n are not near-standard if $n \in {}^*N \backslash N$. We will illustrate the relevance of this concept with two propositions. (The proofs are omitted, see [5] ,pp. 182 and 93.)

<u>Proposition 9.</u> A vector $v_\infty \in {}^*\mathcal{H}$ is near-standard if and only if v is finite and $\sum_{n=m+1}^{\infty} |\langle v,e_n\rangle|^2$ is infinitesimal for all infinite $m \in {}^*N$.

<u>Proposition 1o.</u> A set of points $B \subset \mathcal{H}$ is compact if and only if for every $u \in {}^*B$ there exists a (standard) $v \in B$ such that $u \in \mu(v)$.(The strong topology is presupposed.)

In the last proposition we can replace $\mathcal{H}$ by any topological space. The monad $\mu(v)$ will then be defined in a natural way in terms of the open neighborhoods (loc. cit., p. 9o).

8. LINEAR OPERATORS

The extension of linear operators from $\mathcal{H}$ to $^*\mathcal{H}$ proceeds as for other functions. Thus, if T is bounded and hence defined on all of $\mathcal{H}$, the extension is defined on all of $^*\mathcal{H}$. If T is unbounded and defined on $D \subset \mathcal{H}$, there is a natural extension to $^*D$. However, in general there will be no natural extension to $\mathcal{H}$.

We mention two examples relating to linear operators on $^*\mathcal{H}$.

First: Nonstandard Hilbert spaces allow the construction of eigenvectors which can be associated with continuous spectrum. The following is adapted from [8]:

Proposition 11. Let A be a self-adjoint operator defined on $D \subseteq \mathcal{H}$, and let λ be in its spectrum (continuous or discrete). Then there exists $v \in {}^*D$ such that

$$||Av - \lambda v|| \simeq 0, \qquad ||v|| = 1. \tag{39}$$

We recall that rigged Hilbert spaces offer an alternative approach to such eigenvectors (see e.g. [9]).

Second: We consider a standard Hilbert space $\mathcal{H}$ with a complete orthonormal basis $e_1, e_2, \ldots$, and orthogonal projections P_k, the latter having as range the linear span of $e_1, \ldots, e_k$. The following is well known:

Proposition 12 (standard). Let K be a compact (linear) operator on $\mathcal{H}$. Then for $\forall \varepsilon > 0, \exists\, n \in N$ such that

$$||K - P_n K|| < \varepsilon . \tag{4o}$$

We now sketch two proofs of this proposition, one employing nonstandard notions and one not.

For the nonstandard proof, we make the nonstandard extensions and use the two propositions of §7 (see [5], p. 185). Let $v \in \mathcal{H}$. Then Kv is near-standard, so $||(1-P_m)Kv||$ and also $||(1-P_m)K||$ are infinitesimal for all infinite m. But there is a least integer n satisfying (4o), and so it must be finite.

A proof that uses only standard concepts can be extracted from [1o]. If the proposition were false, then we could find a sequence $\{w_n\}$ of vectors such that $Kw_n \to \overline{w}$ and

$$||(1-P_n)Kw_n|| > \varepsilon \text{ and } ||w_n|| = 1, \ \forall n . \tag{41}$$

One then writes $Kw_n = Kw_n - \overline{w} + \overline{w}$ and makes an obvious use of the triangle inequality to obtain a contradiction.

We are tempted to say that the nonstandard proof is more streamlined, or more direct, even though it uses less familiar notions.

9. THE QUANTUM FIELD MODEL $P(\psi)_2$

The discussion that follows is adapted from that of [2]. We consider a polynomial self-coupling of a scalar field in two dimensions, $P(\psi)_2$, about which we assume that a cutoff theory will have a unique ground state. E.g., the coupling ψ_2^4 has this property.

We first recall some of the steps of the usual Hamiltonian approach, as given in [11] for ψ_2^4. One introduces a family of cutoff functions of the form ($n \in N$),

$$g_n(x) = g(x/n) \quad \text{where} \quad g \in \mathcal{C}_o^\infty,\ g(x) = 1 \text{ for } |x| \leq 1, \tag{42}$$

and the corresponding cutoff Hamiltonians,

$$H_n = H^{(o)} + \lambda \int_\infty^\infty dx\ :P(\psi):(x) g_n(x) + C_n\ , \tag{43}$$

where $H^{(o)}$ is the free Hamiltonian, and $C_n \in R^1$ is least such that $H_n \geq o$. The field ψ acts on the space $\boldsymbol{H}_{\Phi OK}$. By hypothesis, H_n has a unique state $\Omega_n \in \boldsymbol{H}_{\Phi OK}$, satisfying $H_n \Omega_n = o$. We consider

$$\alpha_n(e^{i\psi(f)}) = \langle \Omega_n, e^{i\psi(f)} \Omega_n \rangle\ ,\ f \in \mathcal{C}_o^\infty\ . \tag{44}$$

We have here a state on the C^*algebra generated by the operators $e^{i\psi(f)}$. The space $\boldsymbol{E}$ of states is a compact topological space for the w*-topology.

We can therefore extract a subsequence α_{k_n} of the α_n such that $\alpha_{k_n} \to \alpha_\infty$ as $n \to \infty$. The ΓHS construction then yields a Hilbert space with a representation Π and a vector Ω_∞, satisfying

$$\alpha_\infty(e^{i\psi(f)}) = \langle \Omega_\infty\ ,\ \Pi(e^{i\psi(f)})\ \Omega_\infty \rangle\ . \tag{45}$$

(One can modify this procedure by including spatial averaging which increases with n.Then the final theory will necessarily be translationally invariant. We will ignore this aspect of the problem. We will similarly ignore other properties of the representation Π.)

The following nonstandard construction parallels

the preceeding closely. In place of $\Pi(e^{i\psi(f)})$ we use simply $*(e^{i\psi(f)})$, and we replace α_∞ and Ω_∞ respectively by

$$\alpha = [(\alpha_1, \alpha_2, \ldots)] \quad \text{and} \quad \Omega = [(\Omega_1, \Omega_2, \ldots)] \quad , \tag{46a}$$

so that

$$\alpha(e^{i\psi(f)}) = \langle \Omega, {}^*(e^{i\psi(f)})\, \Omega \rangle \; . \tag{46b}$$

We see here a way to realize the ΓHS construction in ${}^*\mathbf{H}_{\Phi OK}$.

We now complete the construction in the obvious way. We form finite linear combinations of vectors $^*(e^{i\psi(f)})\Omega$, with $f \in \mathbf{C}_0^\infty$, we discard vectors of infinite norm, and we take equivalence classes of vectors, two vectors being equivalent if they differ by an infinitesimal. (This last step corresponds to taking the quotient space, in the usual procedure.) The resulting space then has to be completed.

Observe also the following: In view of the compactness of $\mathbf{\Sigma}$, $\alpha \in {}^*\mathbf{\Sigma}$ is near standard (Proposition 1o and the subsequent remark). Thus, after taking equivalence classes, we are led to a standard state $\bar{\alpha} \in \mathbf{E}$. Both $\bar{\alpha}$ and α_∞ are limit points of set $\{\alpha_1, \alpha_2, \ldots\}$, but they need not to be the same. We may remove this defect by considering all subsequences of $\{\alpha_n\}$, and then each method will yield all limit points.

1o. QUANTIZED FIELD WITH A POINT SOURCE

We give here a n.s. treatment of the quantized field ψ satisfying the familiar equation (with $m > o$),

$$(\Box - m^2)\,\psi(t, \underset{\sim}{x}) = -\lambda\,\delta(\underset{\sim}{x}). \tag{47}$$

This equation has been much discussed, and in particular, a modified equation was analyzed in [2] using n.s. notions. We are bringing up this model once more, since it will lead to a connection between infinite tensor products and nonstandard Hilbert spaces.

We recall the following about infinite tensor products: Given a sequence of Hilbert spaces $\mathbf{H}_1, \mathbf{H}_2, \ldots$, the complete tensor product $\Pi^{\otimes}\mathbf{H}_j$ is non-separable, but any product-vector,

$$\Pi^{\otimes} v_j, \quad \text{where} \quad v_j \in \mathbf{H}_j,\ ||v_j|| = 1 \; , \tag{48}$$

determines a separable subspace [12]. We are interested

in the case where $\mathcal{H}_j$ is generated by the symmetric tensor products: $\Omega(o,j)$, $u_j, u_j \otimes_s u_j, \ldots$, and the u_j form an orthonormal basis of $L^2(d^3k)$. (The $\Omega(o,j)$ can be called partial vacua.) The usual $\mathcal{H}_{\Phi OK}$ is determined by the vacuum vector $\Pi^{\otimes}\Omega(o,j)=\Omega_o$.

Now, the Hamiltonian associated with (47) can be expressed as

$$H = \int d^3\underline{k}\,\omega(\underline{k})\ (a^+ + q)\ (\underline{k})\ (a^- + q)\ (\underline{k})\ , \tag{49}$$

where $\omega(\underline{k})=(\underline{k}^2+m^2)^{\frac{1}{2}}$, the $a^{\pm}$ are essentially the Fourier coefficients of ψ, satisfying $[a^-(\underline{k}), a^+(\underline{\ell})] = \delta(\underline{k}-\underline{\ell})$, and

$$q(\underline{k}) = \frac{1}{4}\,\lambda\ |\pi\omega(\underline{k})|^{-\frac{3}{2}}. \tag{5o}$$

Heuristically, we have

$$a^{\pm} + q^{\pm} \sim U a^{\pm} U^{-1},\ U^{-1} H U \sim H^{(o)} = \int d^3\underline{k}\,\omega a^+ a\ , \tag{51}$$

where $H^{(o)}$ is the free Hamiltonian, and

$$U \sim \exp\ (\int d^3\underline{k}\,q(\underline{k})\ (a^- - a^+)(\underline{k}))\ . \tag{52}$$

Then U transforms (heuristically) the vacuum Ω_o into the ground state of the Hamiltonian. However, $q \notin L_2(d^3k)$, and this implies that U is not an operator on $\mathcal{H}_{\Phi OK}$.

We proceed therefore as in §9, but we use a special form of approximate H's and U's. Namely, we exploit the basis $u_1, u_2, \ldots$ of $L^2(d^3\underline{k})$ and set

$$q_n(\underline{k}) = \textstyle\sum_{j=1}^{n}\ u_j(\underline{k})\ \int d^3\underline{p}\,u_j^*(p)\ q\ (\underline{p}). \tag{53}$$

(We may select u_j's which vanish sufficiently rapidly at infinity, so that the integral is defined.) Upon replacing q by q_n in H and in U we obtain operators H_n and U_n on $\mathcal{H}_{\Phi OK}$. The ground state of H_n is then given by

$$\Omega_n = U_n\ \Omega_o = \Omega^{(1)} \otimes \ldots \otimes \Omega^{(n)} \otimes \prod_{j=n+1}^{\infty}{}^{\otimes}\ \Omega^{(o,j)} \tag{54}$$

and $\Omega^{(j)} \in \mathcal{H}_j$ but $\Omega^{(j)} \neq \Omega^{(o,j)}$.

We may now proceed as in §9, and represent the ground state of H as follows,

$$\Omega = [(\Omega_1, \Omega_2, \ldots)]\ . \tag{55}$$

The procedure outlined in §9 now yields a Hilbert space which can be naturally identified with a sector ($\neq \mathcal{H}_{\Phi OK}$) of the infinite tensor product.

REFERENCES

1. P. Blanchard and J. Tarski, to be published.
2. P.J. Kelemen and A. Robinson, J.Math.Phys. 13, 187o and 1875 (1972).
3. A. Ostebee, F. Gambardella and M. Dresden, Phys.Rev. A 13, 878 and J.Math.Phys. 17, 157o (1976).
4. A. Voros, J.Math.Phys. 14, 292 (1973).
5. A. Robinson, Non-standard analysis (North-Holland Publishing Co., Amsterdam, 1966).
6. H.J. Keisler, Elementary calculus: an approach using infinitesimals (experimental version) (Bodgen & Quigley, Inc., Publishers, Terrytown-on-Hudson, New York, 1971).
7. D. Ruelle, Statistical mechanics-rigorous results (W.A. Benjamin Inc., New York, 1969), exercise on p. 69.
8. M.O. Farrukh, J.Math.Phys. 16, 177 (1975).
9. J.P. Antoine, J.Math.Phys. 1o, 53 (1969).
10. F. Riesz and B.Sz.-Nagy, Functional analysis (F.Ungar Publishing Co., New York, 1955), p. 2o4.
11. J. Glimm and A.Jaffe, Phys. Rev. 176, 1945 (1968); Ann.Math. 91, 362 (197o); Acta math. 125, 2o3 (197o).
12. J. von Neumann, Composito Math. 6, 1 (1938).

LIST OF PARTICIPANTS

A. Actor, Theoretische Physik der Universität Dortmund
S. Albeverio, Institute of Mathematics, University of Oslo
M.F.G. Dias d'Almeida, Porto, Portugal
I. Andric, Institut "R. Boskovic", Zagreb
C.M. Vassalo Serra Alves, Porto, Portugal
M.C. Oliveira Amorim, Porto, Portugal
M. Bace, Inst. für Theor. Physik, Universität Heidelberg
R. Baier, Fakultät für Physik, Universität Bielefeld
M.L. Machado Cerqueira Bastos, Vila Nova de Gaia Portugal
K. Baumann, Fakultät für Physik, Universität Bielefeld
H. Behncke, Inst. für Mathematik, Universität Osnabrück
M. Le Bellac, CERN, Theory Division, Geneva
J. Bellissard, CNRS, Marseille
M. Benayoun, Collège de France, Paris
N. Bilic, Inst. "Ruder Boskovic", Zagreb
P. Blanchard, Fakultät für Physik, Universität Bielefeld
F. Bopp, Gesamthochschule Siegen
S.K. Bose, Universität Kaiserslautern
G.C. Branco, The City College of New York
O. Bratteli, CNRS, Marseille
E. Brüning, Fakultät für Physik, Universität Bielefeld
D. Buchholz, II Inst. für Theor. Physik, Hamburg
F.J. Lage Campelo Calheiros, Porto, Portugal
V. Canuto, NASA, New York
P. Caraveo, Università degli Studi di Milano
T. Celik, Hacettepe University, Ankara
Chan Hong Mo, Rutherford Lab., Berkshire
P.S. Collecott, Cambridge University, Cambridge
J. Cleymans, Fakultät für Physik, Universität Bielefeld
M. Daniel, University of Athens
S. Dimopoulos, Enrico Fermi Inst., University of Chicago
A.M. Din, Inst. of Theor. Physics, Göteborg University
J.A. Dixon, CNRS, Marseille

R. Dobbertin, Laboratoire de Physique Théorique, Université de Paris VII
G. Dorfmeister, Fakultät für Physik, Universität Bielefeld
F. Dustmann, Fakultät für Physik, Universität Bielefeld
W. Driessler, Fakultät für Physik, Universität Bielefeld
J.-P. Eckmann, Dépt. de Physique Théor., Université de Genève
G. Eilam, Israel Institute of Technology, Haifa
J. Engels, Fakultät für Physik, Universität Bielefeld
V. Enss, Fakultät für Physik, Universität Bielefeld
W. Ernst, Fakultät für Physik, Universität Bielefeld
E. Etim, Department of Physics, University of Ibadan
D. de Falco, Istituto di Fisica, Università de Salerno
M.M. Coelho Ribeira de Faria, Pôrto, Portugal
F. Fleischer, Fakultät für Physik, Universität Bielefeld
P. Fré, Istituto di Fisica Teorica, Università di Torino
K. Fredenhagen, DESY, Hamburg
S. Fredriksson, The Royal Institute of Technology, Stockholm
A. Frigerio, Università degli Studi di Milano
H. Fritzsch, CERN, Genève
J. Fröhlich, Princeton University
H. Galic, Inst. "Ruder Boskovic", Zagreb
A. Gandolfi, Università degli Studi di Parma, Parma
P. Garbaczewski, Inst. of Theor. Physics, University of Wroclaw, Wroclaw
W.D. Garber, Inst. für Theor. Physik, Universität Göttingen
L. Garrido, Universidad de Barcelona
H.R. Gerhold, Österreichische Akademie der Wissenschaften, Vienna
B. Gidas, University of Washington, Seattle
M.T. Rodriques dos Santos Gonçalves, Pôrto, Portugal
E.H. de Groot, CERN, Geneva
H. Grosse, Inst. für Theor. Physik, Universität Wien
F. Guerra, Istituto di Fisica, Università di Salerno
H. Hahn, Inst. für Theor. Physik A, TU Braunschweig
C.J. Hamer, Department of Applied Mathematics and Physics, University of Cambridge
B. Hasslacher, California Institute of Technology
G. Hegerfeldt, Inst. für Theor. Physik, Universität Göttingen
P. Hertel, Fachbereich Naturwissenschaften, Universität Osnabrück
A.C. Hirshfeld, Theoretische Physik III, Universität Dortmund
L. Hudson, Dept. of Mathematics, University of Nottingham
H. Inagaki, Comitato Nazionale per l'Energia Nucleare, Frascati, Roma

P.D.F. Ion, Institut für Mathematik, Universität Heidelberg, Heidelberg
F. Jegerlehner, Fakultät für Physik, Universität Bielefeld
J. Jersak, Inst. für Theor. Physik der RWTH Aachen
K. Johnson, Massachusetts Inst. of Technology, Cambridge
G. Jona-Lasinio, Università degli Studi, Roma
K. Kajantie, Dept. of Physics, University of Helsinki
S.S. Kanval, I.I.T. Kanpur, Kanpur, India
M. Karowski, FU Berlin Fb. Physik, Berlin
R.K. Kaul, Department of Physics, Delhi, India
M. Kiera, Inst. für Theor. Physik der RWTH Aachen
K. Kinoshita, Dept. of Physics, Kyushu University, Fukuoka
S. Kitakado, Inst. of Physics, University of Tokyo
A. Knoth, Fakultät für Physik, Universität Bielefeld
I. Koch, Dept. of Mathematics, Bedford College, London
J. Kogut, Cornell University, Ithaca, N.Y., USA
K. Koller, DESY, Hamburg
K. Konishi, Rutherford Lab., Chilton, Didcot, Oxford
A. Krzywicki, Université de Paris-Sud, Orsay
F. Kuypers, Fakultät für Physik, Universität Freiburg
P. Landshoff, CERN, Genève
R. Lima, Faculty of Science, University of Porto
P. Leyland, CNRS, Marseille
K. Litwin, Niels Bohr Institute, Kobenhavn
H. Lotsch, Inst. für Theor. Physik, Universität Wien
Y. Loubatières, Dépt. de Physique Mathematique, Université du Languedoc, Montpellier
J. Lukierski, Inst. of Theoretical Physics, University of Wroclaw
L.E. Lundberg, Matematisk Institut, Universitet Kobenhavn
M. Magg, Inst. für Theor. Physik, RWTH Aachen
J. Magnen, Ecole Polytechnique, Paris
S. Mallik, Inst. für Theor. Physik, Universität Bern
W.J. Marciano, Rockefeller University, New York
P. Martin, Dept. of Physics, ETH Lausanne
M. Mebkhout, CNRS, Marseille
D. Miller, Freie Universität Berlin
M. Mizouchi, Okayama College of Science, Okayama-shi, Japan
M.C. de Oliveira Gomes Moreira, Porto, Portugal
O.M. Vaz Moreira, Porto, Portugal
A. Mueller, Institute for Advanced Studies, Princeton
P. Mulders, Inst. for Theor. Physics, Toernooiveld, Nijmegen
H. Narnhofer, Inst. für Theor. Physik, Universität Wien
K. Napiorkowski, Wydzial Fizyki, University of Warsaw
A.T. Ogielski, Inst. of Theor. Physics, University of Wroclaw
M.A. Marques de Oliveira, Porto, Portugal

A. Ostebee, Institute for Theoretical Physics, State University of New York, Stony Brook
R. Page, Centro Atomico Bariloche, Buenos Aires
L.S. Panta, Dept. Math. Physics, University of Birmingham
W. Pesch, Inst. für Theor. Physik, TU Hannover
B. Petersson, Fakultät für Physik, Universität Bielefeld
Ch. Pfister, Theoretische Physik, ETH Zürich
O. Piguet, Max-Planck-Institut für Physik, München
F.M. Pires, Porto, Portugal
H. Pohlmeyer, II Inst. für Theor. Physik, Universität Hamburg
E. Predazzi, Istituto di Fisica Teorica, University of Turin
P.J. Provost, Physique Théorique, Université de Nice
W. Pusz, Faculty of Physics, University of Warsaw
L. O'Raifeartaigh, Institut des Hautes Etudes Scientifiques, Bures-sur-Yvette, France
M. do Ceu Fernandes de sa Ramalho, Porto, Portugal
M. Ramon-Medrano, Universidad de Madrid
J. Randa, Dept. of Physics, University of Manchester
M. Reed, Dept. of Mathematics, Duke University
H. Reeh, Inst. für Theor. Physik, Universität Göttingen
J.E. Roberts, CNRS, Marseille
D.W. Robinson, CNRS, Marseille
M. Romerio, Institut de Physique, Neuchâtel
L. Rytel, Inst. of Theor. Physics, University of Wroclaw
C. Sachrajda, Stanford University SLAC
H. Satz, Fakultät für Physik, Universität Bielefeld
P. Scanzano, Fakultät für Physik, Universität Bielefeld
K. Schilling, Gesamthochschule Wuppertal, Wuppertal
U.E. Schröder, Inst. für Theor. Physik, Universität Frankfurt am Main
W. Schröder, Fakultät für Physik, Universität Bielefeld
F. Schumacher, Fakultät für Physik, Universität Bielefeld
L. Sertorio, Istituto di Fisica, University of Torino
Q. Shafi, Fakultät für Physik, Universität Freiburg
K. Sibold, Max-Planck-Institut für Physik, München
M. da Silva, Porto, Portugal
S.B. Skagerstam, University of St. Andrews, St. Andrews, Scotland
A.W. Smith, Dept. of Applied Mathematics, University of Cambridge
G. Sommer, Fakultät für Physik, Universität Bielefeld
P.P. Srivastava, Centro Brasileiro de Pesquisas Fisicas, Rio de Janeiro
I.O. Stamatescu, Inst. für Theor. Physik, Universität Heidelberg
J.P. Steinhardt, Coral Gables, Florida, USA

O. Steinmann, Fakultät für Physik, Universität Bielefeld
A. Stern, Chicago
P. Stichel, Fakultät für Physik, Universität Bielefeld
A. Stoffel, Inst. für Theor. Physik, TH Aachen
R.F. Streater, Dept. of Mathematics, Bedford College, London
L. Streit, Fakultät für Physik, Universität Bielefeld
R.L. Stuller, Imperial College of Science and Technology, Dept. of Physics, London
E. Suhonen, Dept. of Theor. Physics, University of Oulu
I. Szczyrba, Faculty of Physics, University of Warsaw
J. Tarski, Fakultät für Physik, Universität Bielefeld
P. Tataru-Mihai, Fakultät für Physik, Gesamthochschule Wuppertal
R. Tegen, Universität Hamburg, Hamburg
G. Thomas, Argonne National Laboratory, Argonne
H.-J. Thun, Fachbereich Physik, FU Berlin
A. Tounsi, Physique Théorique, Université de Paris
P. Tsilimigras, Nuclear Research Center, Attiki, Athens
L. Turki, Inst. für Theor. Physik, Universität Karlsruhe
A. Ungkitchanukit, Royal Holloway College, London
H. Uschersohn, TFT, Helsinki
P. di Vecchia, Nordita, Copenhagen
G. Velo, Università degli Studi, Bologna
I. Ventura, Universidade de São Paulo
H. Watanabe, Dept. of Applied Sciences, Kyushu University, Fukuoka
F. Widder, Inst. für Theor. Physik, Universität Graz
G. Wilk, Institute of Nuclear Research, Warsaw
J. Willrodt, FB Mathematik, Gesamthochschule Siegen
M. Winnink, Institut voor Theoretische Natuurkunde, Universiteitskomplex Paddepoel, Groningen
E. de Wolf, Dept. Naturkunde, Universität Antwerpen
S. Woronowicz, Dept. of Mathematics, University of Warsaw
W. Wreszinski, Instituto de Fisica, Universidade de São Paulo
D. Wright, Bedford College, London
B.C. Yunn, Fakultät für Physik, Universität Kaiserslautern
Y. Zarmi, Physics Dept., Weizmann Institute

INDEX

GPSR Compliance
The European Union's (EU) General Product Safety Regulation (GPSR) is a set of rules that requires consumer products to be safe and our obligations to ensure this.

If you have any concerns about our products, you can contact us on

ProductSafety@springernature.com

In case Publisher is established outside the EU, the EU authorized representative is:

Springer Nature Customer Service Center GmbH
Europaplatz 3
69115 Heidelberg, Germany

www.ingramcontent.com/pod-product-compliance
Ingram Content Group UK Ltd.
Pitfield, Milton Keynes, MK11 3LW, UK
UKHW051325070726

13610UKWH00014B/78

* 9 7 8 1 4 6 1 5 8 9 2 5 9 *